Laboratory Scientific
Glassblowing
Advanced Techniques and Glassblowing's Place in History

Laboratory Scientific
Glassblowing
Advanced Techniques and Glassblowing's Place in History

Paul Le Pinnet

British Society of Scientific Glassblowers, UK

 **World Scientific**

NEW JERSEY · LONDON · SINGAPORE · BEIJING · SHANGHAI · HONG KONG · TAIPEI · CHENNAI · TOKYO

Published by

World Scientific Publishing Co. Pte. Ltd.

5 Toh Tuck Link, Singapore 596224

USA office: 27 Warren Street, Suite 401-402, Hackensack, NJ 07601

UK office: 57 Shelton Street, Covent Garden, London WC2H 9HE

British Library Cataloguing-in-Publication Data
A catalogue record for this book is available from the British Library.

LABORATORY SCIENTIFIC GLASSBLOWING
Advanced Techniques and Glassblowing's Place in History

ISBN 978-981-124-787-3 (hardcover)
ISBN 978-981-124-788-0 (ebook for institutions)
ISBN 978-981-124-789-7 (ebook for individuals)

For any available supplementary material, please visit
https://www.worldscientific.com/worldscibooks/10.1142/12575#t=suppl

Desk Editor: Nur Syarfeena Binte Mohd Fauzi

Typeset by Stallion Press
Email: enquiries@stallionpress.com

Paul Le Pinnet M.B.E.

Honorary Member and Fellow of the British Society of Scientific Glassblowers

by a lack of the appropriate skills. Paul's books ensure there is a transfer of knowledge and skill to meet the scientific researchers' requirements.

Robert Mcleod

Chairman of the British Society of Scientific Glassblowers

Acknowledgments

As the COVID-19 virus spreads around the world with people in lockdown and social distancing, the similarity between the situation that Isaac Newton found himself in and society today is striking!

While escaping the plague of 1665–1666, Newton's enormously productive time at Woolsthorpe Manor, Nr. Grantham, Lincolnshire, saw him lay the foundations for his theories on calculus, optics, the laws of motion and gravity, in what is often called his "Annas Mirabilis" or year of wonders.

Likewise, the contributors to this book have endured self-isolation during the threat of this dreadful pandemic but have been able to concentrate their minds in writing a chapter for posterity and the benefit of others.

I hereby record their names with pride:

Robert McLeod F.B.S.S.G. and Chairman of the British Society of Scientific Glassblowers. Scottish Universities Environmental Research Centre.

Terri Adams F.B.S.S.G. Senior Glassblower Oxford University.

Ian Pearson Hon. Member B.S.S.G. & Editor of the *B.S.S.G. Journal*.

Gary Coyne California State University, USA, retired.

Greg Purdy Glass Services Ltd. Hamilton, New Zealand.

William Fludgate Fellow of the British Society of Scientific Glassblowers.

Stephen W Moehr F.B.S.S.G. University of York, Chief Examiner Board of Examiners, retired.

Norbert Zielinski Glassblower Technische Universitat Berlin.

Philip Legge Glassblower/Manager Scientific design Ltd., USA.

Jens Koster Senior Glassblower, Chemistry Dept. University of Hamburg.

Jean Francois Boutry, Head Glassblower, Lycee Polyvalent Dorian, Paris, France, retired.

Peter Schweifel Chairman of Verband Deutscher Glasblaser E.V.

Graham Reed B.A. Master and Fellow B.S.S.G.

Jeremy Bolton of Jepson Bolton & Co Ltd.

Michael J Souza Princeton University, USA.

Phill Murray Churchill Fellow.

Alan Gall Archivist for the Institute of Science and Technology.

Julia Bickerstaff BEng Hons. M.I.E.T. B.S.S.G. H.C.A.

Joseph Gregar Argonne National Laboratories.

Phil Jones Bath University.

Stephanie Kubler Preston. For translating Norbert Zielinski's article from technical German into English, which I believe is by any standard "no mean feat" for which I am profoundly Grateful.

Phil Le Pinnet "Applications Technical Lead Officer" whom I have relied upon to guide me from a pre-computer age into an "Enlightenment". Which in many ways was brought about by the fact that the first book was written within a Research Environment and where I had access to all manner of support, whereas this second book has been written within a domestic setting. I must say that Phil has shown a great deal of patience and tact in dealing with a parent.

Finally, **My Wife Michelle,** who has at all times lent support by way of her vast knowledge of English grammar. Michelle informs me that at all other times she has maintained a purely Supervisory role.

— Paul Le Pinnet

Contents

Introduction

Paul Le Pinnet

Fellow of the British Society of Scientific Glassblowers

My previous book was published in 2017 and titled *Laboratory Scientific Glassblowing: A practical training method.*

The book laid out a systematic method of training scientific glassblowers from the basic knowledge required, then onto the techniques designed to develop the skills necessary, which if practiced conscientiously would lead to a level of competence that in turn would then enable the trainee to be in a position to gain greater knowledge, thence be of use to the scientific community in that the Scientist would then be able to conduct experiments at the laboratory level, being able to see "change" such as liquid–liquid separation, crystals coming out of the solution and colour differences. Change can indicate a chemical reaction, which in turn can then be followed up and investigated further, all this being conducted within a relatively inert material such as a borosilicate glass, which, although it has its limitations, is clear and relatively easy to clean.

The scientific glassblower is there to use his or her skills to this end and can be of help while being able to offer advice regarding the practicality of an item of glassware in its design and construction. In the last chapter of the previous book, I called upon respected scientific glassblowers from around the world to contribute a chapter on a particular piece of glassware that had some significance to the end user which may well have been difficult to make but was most certainly interesting.

My motivation for writing the books has been the fact that the people who trained me are no longer with us and unfortunately their accumulated knowledge and experience disappeared with them. Therefore, I have once again called on Scientific Glassblowers from around the world to contribute a chapter either on an advanced technique or one that they felt was significant to the scientist with whom they had been involved in changing their ideas into something practical. I am aware, that there can be several adaptions during this process, often having a chemist enquire "can you just" alter the angle at which a side arm comes off a reaction flask? This simple request can quite often require a complete re-make of the item again and again until the final design is attained.

One of the most humbling things ever said to me as a Research Scientific Glassblower happened toward the end of my working life. At the time, I was doing work for a company involved in Nano Technology. All their work was conducted in glass vessels. A chemist from the company brought a 10 Litre round bottomed flask for repair because a jointed inlet port had been snapped off. As we filled out the required forms, I explained that the annealing oven was already full and up to annealing temperature. I would have to make the repair the following day. I then proceeded to set the flask in the lathe together with an off-set holder to align the replacement joint into the correct position for the repair in readiness for the following day.

It was at this point that I realized that the chemist had not left the workshop and was watching silently. I asked if there was anything further that I might do?

His reply was stunning and unexpected, he said "when we get your glassware back to the laboratory, we cannot see the joins"!

I must say that I was staggered by what the Chemist said, only then did I fully realize the appreciation of my work by those who used it. This emphasized to me that what I had been doing the previous 50 years was not "just a Job", although some may feel that way. Feedback of this nature confirmed to me the feeling that I have had for many, many years that Scientific Glassblowing has played a unique part in its interaction with the Scientist and therefore the progress of Science.

Chapter 1

Lasers and Lathes

Jeremy Bolton

Owner and Managing Director of Jepson Bolton Ltd.
rmybolton@yahoo.uk

Lathes in glass tubular processing are amazingly adaptable, they enable glass tubes as small as 1mm OD and as large as 1 metre OD to be processed — although not on the same machine, mind you. Flanges formed, tubes joined, threads created, side arms connected, some can be done freehand and others using mandrels either supported on the burner carriage or held in the tailstock.

In the glass industry today, product designers are asking glass-blowing to produce more intricate designs and to tighter tolerances.

Lasers in Glass Processing

Nowadays, we have a great choice of heat sources, which can be selected to exactly match the glass product you wish to process.

The most common heat sources are mixtures of gases such as Natural Gas, Propane, Butane, Compressed Air, Hydrogen or Oxygen, which can be paired and mixed to be delivered via many different types of burners to give you precisely the type of flame you want for the process or part you wish to make, join or seal.

However, these gases do have disadvantages such as possibilities of contamination originating from the gas cylinder, burner or hoses and difficulties in getting the mixture/flame exactly the same from one batch to another, which then affects the end product parameters.

The other source of heat comes from electricity powering heating elements or induction coils, or a more recent development, lasers.

To heat glass with a laser in preparation for automatic forming or joining processes, CO_2 lasers are recommended. They emit radiation in the mid infrared range; a typical CO_2 laser wavelength is 10600nm. Regardless of the kind of glass, it behaves as opaque and not transparent when targeted with this wavelength. Thus, the laser radiation is absorbed and/or reflected at the glass surface. The radiation of the reflection depends on the radiation situation and the angle of incidence and is about 20%. Consequently, 80% of the laser beam is absorbed on the glass surface and is used to heat the glass.

The lasers direction, power and focus can be controlled via mirrors and lenses, which gives the operator precise control over the area to be heated (from a pin prick size to a 20mm+ circle), and with a pyrometer feedback loop, the exact temperature the glass should be heated to as part of the process. Due to the complexities and power of these types of lasers, they are not to be used in a manual operation but match well in automatic glass lathe operations where the laser is mounted above the work piece and it is the work piece that moves back and forth in the lathe.

Automatic lathe systems with which a laser best works consist of a normal glass blowing lathe, but the movement of everything can be controlled via a Stepped Spreadsheet so that the headstock, tailstock, burner and tool carriage plus selection of forming tools are all controllable in speed of movement, distance traveled, position, etc. And once all the steps have been programmed and adjusted by a skilled glassblower, the lathe will repeat the production process exactly and can form the centre of a fully automatic workstation with loading and unloading by robot arm.

Applications for heating via lasers can include laboratory glass forming (borosilicatel and soda glass), Lamp glass (Quartz), Medical Glass.

In an effort to speed up production and achieve better repro-ducibility of glass processing, forming tools can usually operate during heating, which leads to shorter process times. Common materials for forming tools can be grey cast iron, graphite or stainless steel. The material is based upon the forming process and the kind of glass to be formed.

Other advantages of lasers as a heat source are that they can also be used as a cracking off or cutting device — the laser can be highly focused to irradiate the glass completely, thus providing a very clean cut, eliminating the need for processing the glass into cut lengths using diamond saws and all the cleaning and drying that this extra process requires.

Arnold P1040 Laser Workstation

Summary of advantages of Laser Burner processing in an Automatic lathe workstation include the following:

- Ability to perform several fully automated processes in one workstation.
- Focused Cutting and defocused Heating.
- Possibility of whole length tube processing so reducing pre-processing tasks and saving time.

- More control over heating rates.
- No chemical contamination in the tubes due to condensation Force-free flame processing — no flame pressure.
- Flexible programming for many different standard production tasks.
- One Laser can run with two or more work stations due to speed of heating and forming.

Illustration

Example of a laser workstation is the Arnold P1040 Laser Workstation

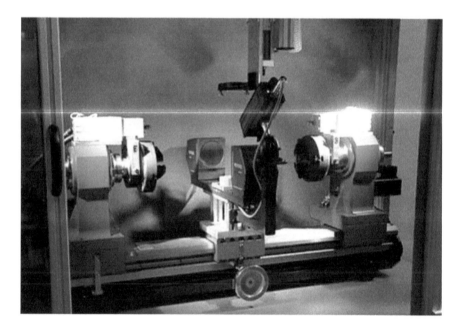

Chapter 2

Joseph Priestley a Man of the "Enlightenment"

Paul Le Pinnet

Fellow of the British Society of Scientific Glassblowers

In 2011, I gave a talk on the history of Scientific Glassblowing in the Science Museum, London. It was part of a series of lectures held throughout the day, but where to start a history? I have always thought that the transition from Alchemy to Chemistry was during the Enlightenment of the late 18th century with experiments being conducted more and more in glass vessels. Since that time Scientific Glassblowing has developed alongside Chemistry.

I have lived most of my life in Warrington and have been aware of the name of Joseph Priestley as he taught at the Warrington Academy (Figure 1).

During the 1760s, this non-conformist academy was described as the "Athens of the North". The Academy attracted free thinkers from around Britain, Europe and America. While there, Priestley met such people as Benjamin Franklin, Matthew Boulton and Josiah Wedgwood. Priestley taught English Grammar and during his spare time conducted Chemistry experiments, there are just three items of his in the Warrington Museum, a glass retort, a vacuum pump and a burning glass used to direct the sun's rays onto various chemicals; While in London, I returned to the Science Museum expecting to see more of his Glassware. I was somewhat disappointed as the only piece belonging to Priestley was his Electricity Generating Machine.

Figure 1.   Warrington academy.

Apparently in 1764 Benjamin Franklin encouraged Priestley to study Electricity rather than chemistry. Two years later Priestley published a seven-hundred-page book titled *The History and Present State of Electricity*. It outlined the conductivity of metals and explained the difference between conductive and non-conductive materials; his text became the standard for the next one hundred years. Michael Faraday used it for his work on electromagnetism; William Hershel's infra-red radiation, Henry Cavendish and James Clark Maxwell both relied on Priestley's work. As a result, Priestley was made a Fellow of the Royal Society.

Priestley continued his Chemistry experiments; he was fascinated with what he described as "common air". He knew that the air we breathe was a gas. The question was is it a single gas or a mixture and how would he be able to prove it either way? He developed his pneumatic bath in which he conducted a series of experiments; the bath consisted of a series of inverted domes within a bowl of water which made a gas-tight seal (Figure 2).

A lit candle will give off both light and heat. A dome of known volume is placed over the candle which continues to burn. As the candle dies, the water rises taking up 21% of the volume no matter

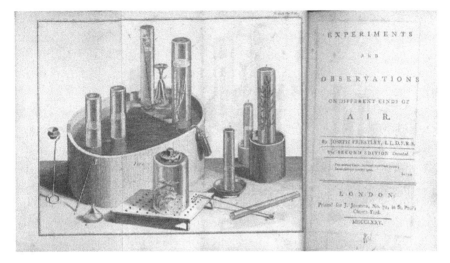

Figure 2.   Common air experiment.

what the shape or size of the dome, proving that air is indeed a mixture. Priestley repeated his experiment with a small caged animal suspended above the water, when this died, the water rose to exactly the same level, he called the gas that supported both combustion and life deflogisticated air, the remaining air he called Nitrous air. We now know that air consists of 78% nitrogen, 21% Oxygen and 1% Argon Inc., traces.

Within his pneumatic bath he also conducted photosynthesis experiments proving that vegetation produces Oxygen!

Priestley burned mercury Oxide. He noted that a candle which was placed nearby it burned with "remarkable vigorousness". In 1767, he moved to Leeds and became a church minister. Priestley lived next to a brewery so had a good supply of Carbon di Oxide. He continued his chemistry as a hobby. In 1773, Lord Shelbourne offered him a post as tutor to his two children, and Priestley moved to Bowood house, Calne near Chippenham. He was given time and money to equip a laboratory. While at Bowood he published a number of books, copies of which are in the Warrington reference library.

Thomas Payne who had written *The rights of Man* was a good friend of Priestley; together they supported the Americans in their war of independence against Britain, 1775–1783. They both also gave their support to the "French Revolution". This brought Priestley into direct conflict with the British Government. Lord Shelburne, on the insistence of his wife, told Priestley to leave but gave him a pension and enough money to buy a house.

Priestley moved to Sparkbrook in Birmingham with his family and his equipment where he set up a laboratory and library. Sadly, the local people did not like Priestley's radical views. In June of 1791, three men tried to enter his house while shooting pistols through his windows, just missing a maid servant. One night in July he was warned that a mob was on its way to "Do him harm".

Priestley and his wife escaped and hid for three days, eventually finding refuge in London. The mob ransacked his house (Figure 3), laboratory and library.

On 19th July 1791, he wrote a letter to the people of the town of Birmingham.

> *"You have destroyed the most truly valuable and useful apparatus. You have destroyed the Library corresponding to that apparatus which no money can re-purchase.*

Figure 3.   Destruction of Priestley's House.

You have destroyed manuscripts which have been the result of laborious study of many years and which I shall never be able to re-compose.

Yours faithfully
J. Priestley"

Two years later Priestley left for America. He lived in Northumberland County, Pennsylvania. Sadly, his son died soon after arriving in America, his wife died one year later. Priestley continued experimenting up to his death in 1804.

The French Chemist Antoine Lavoisier and his wife were contemporaries of Priestley (Figure 4).

Marie Anne Pirette Paulz or "Madam Lavoisier" was well educated, speaking French, English and German, and with a good understanding of Latin. She worked with her husband and wrote up the experiments. Madam Lavoisier translated the works of Priestley, Cavendish and the Swedish Chemist Carl Schiel for her husband to read and made observations in the margins.

Antoine then repeated the experiments; he understood that nothing is created or destroyed. It changes phases, but the mass remains the same. He defined elements for the periodic table and renamed dephlogisticated air as Oxygen! He defined compounds and used quantitative analysis.

His greatest failure was that he claimed the work of others without crediting their work. It was the Royal Society who rectified the situation on behalf of Priestley.

In December 1793, Antoine Lavoisier was arrested, imprisoned, put on trial and condemned. He asked for time to complete an experiment but was told that the Revolution had no need of Chemists or Scientists. On the 8th May 1794, he was taken with 26 others and guillotined.

Before his trial Madam Lavoisier realized that if convicted, his property and goods would have to be forfeited. She gave the notes, drawings and published works to friends for safekeeping. Soon after Antoine's execution, Madam Lavoisier was arrested, imprisoned to await trial and probable execution because she had been married to

Figure 4.   Antoine & Madam Lavoisier.

a convicted traitor. The revolution ended abruptly with the fall of
Maximillian Robespierre. Madam Lavoisier was released but found
that her house had been emptied; friends rallied round and gave her
a table, chairs and a bed to lie on. Some one and a half years after

Antoine's execution all his glassware and laboratory equipment were returned together with a small note saying: to the widow Lavoisier returned the property of Antoine Lavoisier Chemist, "Executed in error".

This was the only apology she ever received; she exhibited the glassware to raise money, as she grew older, she sold off some items. She died in 1836.

The man who accused Antoine Lavoisier, which brought about his death, was Paul Marat. Born in Switzerland and trained as a Doctor in Paris, London and Edinburgh, he was given his Doctorate by St. Andrews University. Paul Marat spent time at the Warrington Academy; he worked with Priestley on his early experiments.

Marat returned to Paris and became a ferocious leader of the Revolution. He called for the death of the King, his family, all aristocrats and landowners. Lavoisier was on his list because he had derided Marat's skills both as a Doctor and Scientist.

The Committee of Public Safety refused to condemn Lavoisier as he oversaw the production of gunpowder for both the Army and Navy. But they would keep his name on file! Paul Marat was assassinated 13th July 1793 (Figure 5), by Charlotte Corday who thought that without Marat the terror of the Revolution would end, but it grew worse.

I am delighted to say that all of Madam Lavoisier's engravings, notes and published works plus most of the glassware is safe and well in Cornell University, U.S.A. What better place for it to reside as France rejected Lavoisier and Britain rejected Priestley?

Eventually the Royal Society acknowledged that Priestley worked for the good of Science and gave him the Copley Medal. Other recipients are Benjamin Franklin, Humphrey Davy, Michael Faraday, Justus Von Liebig, Robert Bunsen, Ernest Rutherford, Albert Einstein and in 2006 Steven Hawkins.

Joseph Priestley is in good company.

Finally, an engraving by Madam Lavoisier (Figure 6), without whom her husbands and Priestley's work would have disappeared.

Figure 5.   Marat's murder.

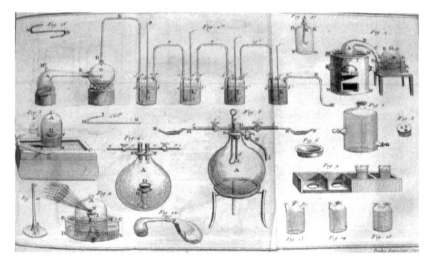

Figure 6.   Engraving by Madam Lavoisier.

Look at the cylindrical vessels and imagine them to be Spherical with interchangeable joints and we can think of a gas transfer manifold complete with manometers and associated paraphernalia, which looks to me very much like modern Scientific Glassblowing.

Chapter 3

Lycee Dorien, Glass Training School, Paris, France

Jean-François Boutry

In 1931, how amazing, a scientific glassblowing school was created. Could it be the first one, no idea!

LYCEE POLYVALENT DORIAN
Technologique – Scientifique -Professionnel
74, avenue Philippe Auguste
75011 PARIS France

DORIAN TECHNOLOGICAL, VOCATIONAL AND SCIENTIFIC SCHOOL

Scientific glassblowing School

☎ : 00-33-(0)1-44-93-81-30
🖷 : 00-33-(0)1-43-79-10-72
e. mail : <u>verre.dorian@wanadoo.fr</u>

ENGLISH   Scientific glass blowing   FRANCAIS
☎: **00-33-(0)1-44-93-82-70**

The school Dorian was built in 1880 by Caroline Holtzer, wife and widow of Pierre-Frederic DORIAN Minister of the 3rd Republic. The aim at that time concerned the war orphans and already some vocational training was being given about wood and metal works.

Amazingly, a scientific glassblowing school was created in PARIS in 1931.

School Dorian 1910

Scientific Glassblowers Students in 1958

Presentation of the Scientific Glassblowing Department and the Students Programme

The "Lycée Régional Polyvalent DORIAN" (college) offers for over the last 90 years scientific glassblowing courses via the education authority, which means that it is absolutely free for the students, which is rather unusual according to the cost of the training in terms of glass and energy.

The "Lycée Régional Polyvalent DORIAN" has two levels of general studies, which accompany the vocational trainings:

- The CAP (Certificat d'Aptitude Professionnel) close to a "GCSE" level
- The "Baccalauréat Professionnel" close to an "A" level.

Taking into account that part of the timetable includes vocational training, the level of the academic education in these two exams is a bit lower.

Some students are taking the "CAP" level course over two years' training in scientific glassblowing.

Some other students take the "Baccalauréat Professionnel" level course over three years in scientific glassblowing. This form was European from 2008 till 2013. It meant that the students received two hours more of English per week, one of which was completely devoted to scientific glassblowing. This European class can only go on if there is a scientific glassblowing teacher whose level in English is very good. Luckily, a new scientific glassblowing teacher is joining the school so it should start again very soon. It is very important as it helps some students apply for a vacancy abroad once their training is completed.

A student who passed the "CAP" level can be integrated into the 2nd year of the "Baccalauréat Professionnel" level course and will complete all four years of studies.

It was in 1957, at the request of research engineers, professors and professional organizations, that it was decided to set up the "Baccalauréat Professionnel" level for glassblowers, from then on they would be called technicians. This "Baccalauréat Professionnel" level enables them to have a better understanding of what is required and above all to become involved in industrial management and in the research field.

Industrial managers and scientific glassblowers from universities and research centers meet together, along with the teachers. Their aim is to set up educational programs, enabling the students to train for a professional life within the duration of the course. These meetings are of course under the control of the educational authorities' inspectors. The presence of a Professional Organization and the scientific glassblowers from universities is indispensable as it orientates the training and above all justifies the necessity for around 15 new jobs a year mainly in France and there are quite a few opportunities around the world.

Thanks to the good relationship between all the actors involved in the training, all of the students have a job once their studies are completed, at either level of exam sat. The main jobs concern borosilicate and silicate glassblowing and glass–metal sealing.

The teachers attribute the jobs according to the positions available. Around 12 students leave the school having successfully completed their education every year.

Some students on leaving have gone on to work abroad (Africa, Belgium, Canada, Columbia, Germany, Sweden, Switzerland, UK, USA, etc...). These companies from abroad have told the school that they appreciate the seriousness of the training which is dispensed. Nevertheless, the school is quite aware that further progress needs to be made by the students on their arrival in the professional world in terms of quality and quickness.

The companies enable this training to be carried out due to an apprentice tax that they pay to schools according to French law. The annual subventions from the educational authority alone wouldn't be sufficient and investments need to be regularly made for this training.

At the moment, in the scientific glassblowing section, there are 24 work benches, equipped with the necessary torches, 11 equipped lathes, 1 oven, 1 drilling machine, 1 grinding machine, 1 engraver,

1 vacuum equipment, a glass-cutting machine, 5 computers, etc... Borosilicate-glass, soda-glass, silica-glass, glass–metal sealing with glass chain and laboratory glass components are used or made by the students.

Some students are given a basic approach to artistic glass so as to enlarge the glass applications to involve two additional ovens, sanding machine, engraving machine and the use of coloured glass.

In France, a school year represents 34 weeks (52–18 weeks of holidays).

Each student does on average 27 weeks' training a year at school, which includes per week approximately 12 hours of practical exercises, (in scientific glassblowing, 70% for the bench and 30% for the lathe, see pages further on showing some of the practical exercises done by students the 1st, 2nd, 3rd and sometimes 4th year), 1 hour of scientific glassblowing technology, 2 hours of technical drawing, 3 hours of artistic education, 2 hours of Sport and 14 hours of general studies (Maths, Physics, Chemistry, English, French Literature, Civil Education and Business Management, Hygiene and Safety). Some of the hours can slightly change along the years.

The students who passed their "CAP" level and who do not wish to continue can do so. However, some of them carry on for two more years to take the "Baccalauréat Professionnel" level to get a higher general education. It has a great importance within a company or when applying for a job in a university. This "Baccalauréat Professionnel" level diploma is really very important for our profession.

At the same time as the Baccalauréat Professionnel exam, a challenge called "best apprenticeship in France" was created 15 years ago. The best students of the school in their last year studying and the apprentices doing their whole training in a company are concerned. See at the end of this topic is one of the gold medalists "chef d'oeuvre" done by two students from the school.

These days around 90% of French scientific glassblowers have this basic training in this school. Many of them phone, write or come back to visit the school. There is always a permanent contact, which is very beneficial both to the new students and to the former ones. These relations are very important because the students spend on average seven weeks (35 hours/week) per year on each level of exam within a company or a university where the glassblower in charge is a former student himself. Some students may also have a course abroad (in England and Italy through the Erasmus Program) to learn more and to demonstrate their capabilities. These courses are added to the 27 weeks at school.

To resume, the number of practical training hours during a year is approximately 569 hours.

Once a year there is an open day at the school and this enables us to select students at the age of about 15–6; there is a participation at employment forums and invitations for future students to mini-courses to see what it is all about. Sometimes a few are sent to us because they have already a contact with the glassblowing industry.

Within this school there is a will to be open and to show what is going on in our profession in the rest of Europe. So, when it's possible, school trips are organized in Europe. Over the last few years, the students have been taken, sometimes twice or three times, to Belgium, Czech Republic, Denmark, England, Germany, Holland, Italy and Sweden.

The purpose is to give more than a complete professional training to these students. The aim is that their general education must be fulfilled so that they may accomplish to their best ability the speciality that they have chosen or if necessary that they reorient their views to something else. Should this be the case, then, undoubtedly the general education that they have received within the scientific glassblowing program should enable them to look elsewhere in case of reorientation.

Jean-François Boutry (retired teacher from the school DORIAN, scientific glassblower in England for a while and BSSG member since many years).

Some practical exercises (1st year "CAP" and "Baccalauréat Professionnel" level)

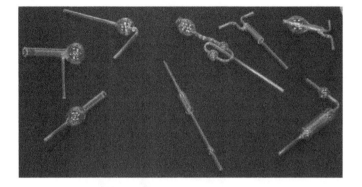

(2nd and final year "CAP" and 2nd year "Baccalauréat Professionnel" level)

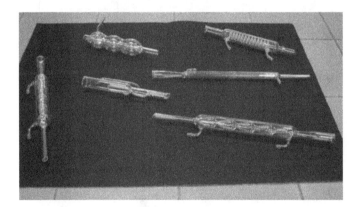

Some practical exercises (3rd and final year "Baccalauréat Professionnel" level)

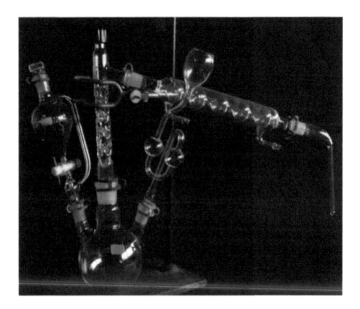

Some practical exercises (3rd and final year "Baccalauréat Professionnel" level)

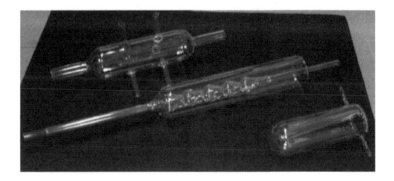

Some practical exercises (3rd and final year "Baccalauréat Professionnel" level)

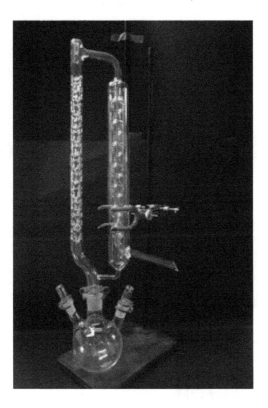

Some practical exercises (3rd and final year "Baccalauréat
Professionnel" level)

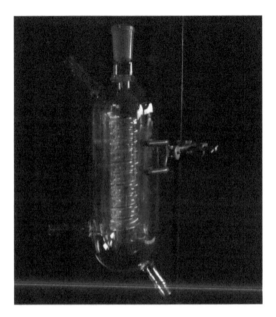

In their final year, a few of the best students attempt to win a
competition called "The best apprentice in France" This competition
is open to anyone under the age of 25, learning the trade in the
scientific glassblowing field.

The following apparatus belongs to two students from the school who won gold medals in 2012.

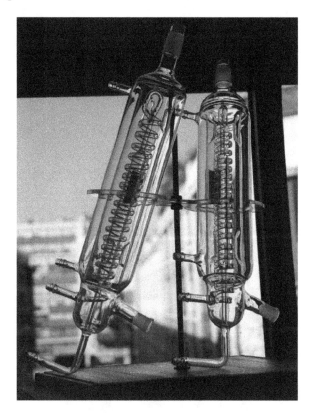

The following apparatus belongs to a student from the school, who won the gold medal in 2013.

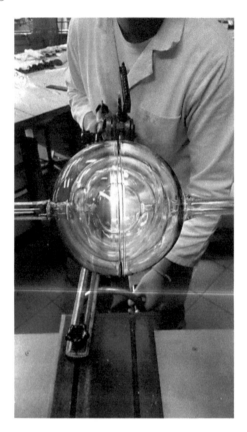

Chapter 4

Where Scientific Glassblowing Is Trending and How Can We Adapt?

Michael J. Souza

Princeton University, USA

This chapter proposes the thesis that in the near future, if we want to enhance our earning potential, scientific glassblowing has to transform from a highly skilled manufacturing base and into a specialized service industry for new research and technology. During this transformation, being a member of the ASGS will play a substantial role in one's success.

Preface

The opinions expressed here by me are based on 40 years' experience as a scientific glassblower. Nearly half of my time was spent working in private industry and the rest as the glassblower at Princeton University. So while it is said that, "To a hammer, everything looks like a nail..." I will try and support this thesis by looking at the past and describing trends that indicate there may be some basis for my hypothesis.

The golden age of scientific glassblowing: 1940s–1970s

Beginning in the 1940s, the world was at war, and in its aftermath, the United States was the lone economy that had not been ravaged as a result of World War II. What followed was a longer, more abstract conflict known as the Cold War. The battlefield was not as much between soldiers as it was for scientific and technological advantages.

By the 1940s, **electronic glassware** was an enormous source of jobs in scientific and industrial glass manufacturing. This field had ushered in new technologies such as radar, microwave, television and computers. All of these components required glass vacuum tubes. The job was tedious as most of the tubes were fabricated in assembly line fashion. However, the fallibility of the tubes (due primarily to burnt out filaments) and their demand assured the industry plenty of work.

A case in point is the ENIAC (Electronic Numerical Integrator and Computer). ENIAC was designed to calculate artillery-firing tables for the United States Army's Ballistic Research Laboratory and was first used to calculate statistics for modeling the hydrogen bomb.[1] It contained 17,468 vacuum tubes, 7,200 crystal diodes, 1,500 relays, 70,000 resistors, 10,000 capacitors and around 5 million hand-soldered joints. It weighed more than 30 tons and took up more than 1,800 square feet.[2]

Several tubes burned out almost every day, leaving it non-functional about half the time. Special high-reliability tubes were not available until 1948. Most of these failures, however, occurred during the warm-up and cool-down periods when the tube heaters and cathodes were under the most thermal stress. Engineers reduced ENIAC's tube failures to the more acceptable rate of one tube every two days. According to a 1989 interview with Eckert, "We had a tube fail about every two days and we could locate the problem within 15 minutes."[3] In 1954, the longest continuous period of operation without a failure was 116 hours — close to five days.

---

[1] ENIAC's first use was in calculations for the hydrogen bomb. Moye, William T. (January 1996). "ENIAC: The Army-Sponsored Revolution." US Army Research Laboratory. Accessed July 9, 2009.

[2] http://encyclopedia2.thefreedictionary.com/ENIAC. Accessed on May 19, 2013.

[3] Alexander Randall 5th "A Lost Interview With ENIAC Co-inventor J. Presper Echert." *Computer World* (February 14, 2006): 3. http://www.computerworld.com/s/article/108568/Q_A_A_lost_interview_with_ENIAC_co_inventor_J._Presper_Eckert?taxonomyId=12&pageNumber=3. Accessed May 19, 2013.

Original photo courtesy of the curator, released under GNU license. Copyright 2005 Paul W. Shaffer, curator of the University of Pennsylvania ENIAC Museum.

Photo shows the ENIAC. Photo courtesy of U.S. Army.

Detail of the back of a panel of ENIAC, showing vacuum tubes. This was taken from the computer lab, which has a glass window to the back of the piece of ENIAC on display at the Moore School of Engineering and Applied Science, at the University of Pennsylvania.

Fifty Years Later

By the 1960s, the vacuum tube was quickly being replaced by small transistors about the size of miniature jellybeans. They issued in new products like the transistor radio and at the same time ushered out a whole segment of glass industry jobs in electronic glassware. By the 1970s, electronics were reinvented once more as the age of semiconductors quickly replaced the diode transistor. And on the 50th birthday of the invention of ENIAC, the University of Pennsylvania recreated the mammoth-sized computer with all of its functions on a silicon chip. This chip would be smaller than a postage stamp measuring just 7.44 by 5.29 sq. mm, a chip which uses a 0.5 micrometer CMOS technology.

The salient point is that research and displacement technology had evolved in just 20 years from the vacuum tube to the transistor and next to the semiconductor. In our small niche of this universe, glassblowers were first essential in the development of the vacuum tube.

ENIAC on a Chip, University of Pennsylvania, USA, 1995. *In celebration of ENIAC's 50th Anniversary, the machine was reimplemented using modern integrated circuit technology. The room-sized computer could now fit in the palm of your hand. Gift of Jan Van Der Spiegel and the University of Pennsylvania, 102719916.*

However, with the advent of the transistors, our skills were quickly irrelevant. Yet, in the decade that followed, semiconductors proved to be a more disruptive technology. As a result, it created an entirely new demand for scientific glassblowers. Unlike the previous transistors, the ones grown on semiconductor chips are not products that can be stamped out and welded on assembly lines. Instead, chips are grown and hatched in pure quartz glassware.

Indeed this new industry now required a more proficient type of glassblower. One who could fabricate quartz glass to exacting tolerances and build the furnace tubes, boats, carriers and all of the associated infrastructure required to process semiconductors. Those low-paying assembly line jobs on stem presses were replaced by highly skilled and well-paid glassblowers who could fabricate quartzware. More importantly, the age of the digital revolution was born.

Advancements in Chemistry

Since the days of Alchemy, glass and chemistry seem to have a natural affinity with each other, like a hand and glove. Fifty years ago, it seemed every major university with a chemistry department employed at least one or more scientific glassblowers in their department just to sustain all of the glassware required for the laboratories and the research groups. In addition, most chemistry doctoral programs required in-house glassblowers to teach scientific glassblowing to graduate students. Large pharmaceutical companies and chemical companies not only employed but also often groomed their own staff of scientific glassblowers.

That is no longer the case. The world has gotten smaller in many ways: the transportation of goods and services has centralized production. Trade barriers are no longer an issue and products can be shipped to most any location overnight. The digital age with Google, email, faxes, smart-phones and videos connect the user to the provider instantly. So if it can be out on a shelf, it can be purchased when it is required. That means less inventory requirements for the customer. Instead, it puts the cost onto the supplier. And when technology switches quicker than ever, that is an important efficiency. None of this of course is unique to our industry.

NMr Machine.

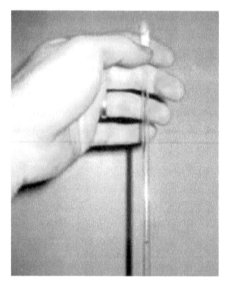

NMr Tube. While the machine itself is often massive, the glassware required is miniscule.

Photos and graphics courtesy of Wikipedia.

However technological advancements in analytical chemistry and in chemistry research as a whole have radically reduced both the size and scope of glassware required in industrial, pharmaceutical and academic chemistry labs around the world.

The foremost change has come from a technique known as **NMr (Nuclear Magnetic resonance) Spectroscopy**, a technique that exploits the magnetic properties of certain atomic nuclei. It determines the physical and chemical properties of atoms or the molecules in which they are contained.

This single advancement now reduced the product requirement to test and characterize new compounds to just a few teardrops (about 1 inch of sample inside a 5 mm test tube). Traditional "wet chemistry" downsized into "minimumware" or "microware" kits. As a result, chemists can begin in volumes as small as 50–100 cc flasks. No longer would labs require liters of materials to distill, fractionate or purify in a tedious and expensive research process. Now that glassware could become smaller, it became cheaper and more easily replaced as new.

Chromatography is not a new technique. It is both preparative and analytical. It separates the components of a mixture and can be used for analysis, or purification for use in a new compound. The premise is to create a mobile phase by one method and transfer it to a structure or material to maintain the component in a stationary phase. It is a productive and elegant technique commonly used in organic chemistry to synthesize new drugs and chemicals. In most cases, the glassware required for this process is primarily a lengthy test tube also known as a separatory column.

However, in the past 20 years due to the use of software and new technology, we have seen many new variations added to this technique that require little glassware. Indeed, presently the following forms of chromatography are available to researchers:

- Liquid
- Gas
- Paper
- Planar Thin Layer,
- Chiral
- Fast Protein
- Counter Current
- Two-Dimensional
- Reversed Phase
- Expanded Bed
- Size Exclusion
- Ion Exchange
- Super Critical
- Affinity, etc.

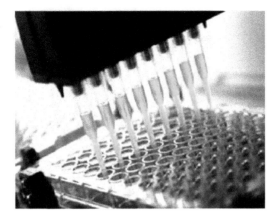

Combinatorial chemistry

Another advancement was Combinatorial Chemistry, which is a sophisticated set of techniques used to synthesize, purify, analyze and screen large numbers of Chemical compounds, far faster and cheaper than was previously possible. Whereas classical synthetic chemistry involves the stepwise synthesis and purification of a single compound at a time, combinatorial chemistry makes it possible to synthesize thousands of different molecules in a relatively short amount of time. The glassware is reduced to slides and miniature vials arranged in trays.

The Spin-Offs and the Patents

While it often seems the primary goal of research is to make one's task in life to be obsolete (after all find a cure for cancer and you are out of work), the outcome is more likely to produce a spin-off. In the field of traditional chemistry, these rapid advancements have narrowed the requirement for scientific glassblowers in the chemistry departments at universities and at pharmaceutical companies. They no longer need their own in-house manufacturing to run their labs. Instead, the field of inquiry has broadened. Indeed, the broadening of chemistry research has ventured into and created whole new areas of research.

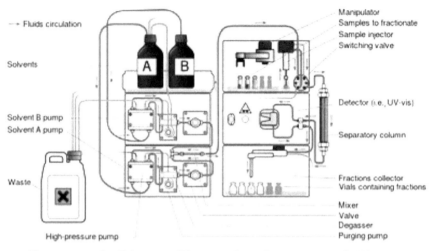

Chromatography Schematic. Photos and graphics courtesy of Wikipedia.

For instance, organic chemistry has found ways to make carbon-based polymers and small molecules with incredible electronic properties. They can emit light, store photovoltaic energy and can even be used in lasers and transistors. Is it chemistry, or is it a materials science, electrical engineering or photonics? The answer is yes to all of those disciplines. Actually, all of these fields now fit into a category known as "Applied Research" and its roots can be traced to a very important piece of legislation known as "The **Bayh–Dole Act** or **Patent and Trademark Law Amendments** December 12, 1980."

This uniquely bi-partisan piece of legislation provided key changes in the ownership of inventions created from the use of federal funding. Before the Bayh–Dole Act, federal research funding contracts and grants obligated inventors (wherever they worked) to assign inventions they made using federal funding to the federal government.[4] Instead Bayh–Dole permits a university, a small business or a non-profit institution to elect to pursue ownership of an invention

---

[4]Stevens, A., "The Enactment of Bayh–Dole," *Journal of Technology Transfer*, 29: 93–99 (2004).

in preference to the government. Prior to the enactment of Bayh–Dole, the U.S. government had accumulated 28,000 patents, but fewer than 5% of those patents were commercially licensed.[5]

As a result,

> Universities and their inventors earned more than $1.8-billion from commercializing their academic research in the 2011 fiscal year, collecting royalties from new breeds of wheat, from a new drug for the treatment of HIV, and from longstanding arrangements over enduring products like Gatorade.
>
> Northwestern University earned the most of any institution reporting, with more than $191 million in licensing income.... The 617 start-up companies formed in 2011 were a slight increase from the 613 reported in the previous year. Start-up companies appeared to be a growing focus for some of the institutions in the survey. In 2010, 12 institutions reported forming 10 or more companies; in 2011, 14 institutions did so....
>
> The 157 universities that responded to the annual survey of the Association of University Technology Managers, released on Monday, completed 5,398 licenses and filed for 12,090 new patents. They also created 617 start-up companies.[6]

Fruits of The Golden Age of Scientific Glassblowing: New Sciences, Innovations and Incentives

As I have tried to illustrate, the pace of technology, the lowering of tariffs and globalization has transformed many of our jobs from a narrow manufacturing base of vacuum electronics and industrial chemical glassware suppliers to a more service-based industry of "Life Sciences" and specialty shops that can flourish if they are adaptable and can utilize special talents to address customers' needs. This trend will continue to change the landscape of our work. One can easily list

---

[5] "Technology Transfer, Administration of the Bayh–Dole Ace by Research Universities," U.S. Government Accounting Office (GAO) Report to Congressional Committies. May 7, 1978.

[6] Blumenstyk, G., "Universities Report $1.8 ~Billion in Earnings on Inventions in 2011," *The Chronicle of Higher Education*, (August 28, 2012). http://chronicle.com/article/University-Inventions-Earned/133972.

the following new technologies developed in the last few decades in which scientific glassblowing can mine new fields of work:

- **Photonics:** The manipulation of light, i.e., lasers, emission, transmission, modulation, signal processing, amplification and detection or sensing of light.
- **Organic electronics:** OLEDs, plastic electronic materials, carbon based.
- **Environmental sciences:** Physical and biological, atmospheric, soil science.
- **Alternative energy:** Fuel cells, battery research, CO_2 reduction, bio-fuels, solar energy.
- **Nanotechnology:** MicroElectronicalMechanicalSystems (MEMs), carbon nanotubes, graphene, sensors.
- **Molecular biology:** Biological processes, DNA, RNA, cloning, stem cell research.
- **Bio-medical:** Vascular models (Farlow Scientific), diagnostics, transplants, new funding to map brain function, artificial limbs and sensors.
- **Government expansion:** NASA, Dept. of Energy, NIST, EPA, DARPA.
- **The Horizon:** Sspintronics, quantum computing, zero gravity materials, dark matter and dark energy, neutron filters and **Serf Magnetometry**.

There are also new techniques available:

- **Bonding techniques:** Diffusion, anodic, transfer tapes, frittable glasses, brazing and epoxies.
- **Machining:** Water-jet cutting, laser, ultra-sonic, CNC, centerless-grinding, lapping, polishing, etc.
- **New glasses and ceramics:** Sealing glasses, optical glasses, envelope glasses, sapphire, sol-gel coatings, etc.

The ASGS, More Important THAN Ever

We randomly picked 20 famous experiments that changed our world — Thomson's discovery of electrons, Faraday's work on electricity and Newton's splitting of white light into its component colors with a prism, for example — and found that 15 of them could not have been performed without glass tools.[7] Alan MacFarlane, Department of Social Anthropology, University of Cambridge.

Science and technology continuously transform our Society and the workplace. Indeed, the jobs for making buggy whips a century ago were not so much lost as they were transformed into jobs in the new auto industry. While glass continues to be an indispensable tool in shaping our future, we must continue to adapt. The best way to do this is by becoming an active member of the ASGS:

- If you look through the archives of our publications (**Fusion** and **Proceedings**), you can easily see the chronicles of advancements taking place, both historically and in real time.
- If you participate in our electronic **ASGS bulletin board**, attend **Section meetings**, you are plugged into a network where problems can be solved, key suppliers can be identified and helping hands are always extended.
- If you go to **symposia**, attend **professional seminars, workshops and technical papers**, you will get a deeper understanding of the "science" behind scientific glassblowing.
- **If you participate as an officer, a committee chair**, contribute to publications, technical posters and workshop demonstrations, you can build a resume filled with accomplishments. More importantly you will have the satisfaction of being part of a legacy and a mission that truly elevates our profession.

In the near and distant future it is easy to suggest that technologies will grow at a faster pace. However, our recent history indicates that there are two ways to compete for those jobs as scientific glassblowers. One way is by manufacturing glass at lower margins, tighter efficiency and less pay. Even so, this downward spiral still does not address automation.

---

[7] MacFarlane, A. and Martin, G., "Beyond the Ivory Tower: A World of Glass," *Science,* 305: 5689 (September 3, 2004), 407–1408. DOI: 10.1126/science.109397.

The alternative way is to become more intimate with those technologies and find ways to service their needs by being more informed and more adaptable in the marketplace. In this future, ideas and timely service will be a premium commodity. To succeed in this venue, you will need a deeper understanding of a customer's process and partner with them in the creative process. The pertinent question you should ask is, "What do you want your glassware to do?" and not simply, "What do you want me to build?" And the answers to those questions can best be found by being an active member of the ASGS.

Chapter 5

Stealth & Hasty

William Fludgate

To those in the glass industry, the title might be more commonly known as Health & Safety (H&S). Yes, it is absolutely paramount that we must follow all the guidelines to promote a safe environment to continue our best work practices to the utmost. So why have I called it as such. Well, when you consider that at one time, I had an insurance assessor visit the workshop and be mortified that we were putting glass into our mouths and placing our hands so close to flames that were in front of our faces, let alone snap glass tubing with our hands after drawing a scratch across the long length of tubing. He insisted that we had a face shield covering our face, despite the fact we wore our tinted safety glasses. He wanted a shield over the bench burner, plus we had to engage a machine to cut glass tubing. To then insist that we had to wear heat proof gloves was just laughable. His report was petty damning to say the least; he would shut down our industry in a heartbeat. He did not get our business.

"So why stealth and hasty", this was put to me as a young apprentice, full of anticipation on my road to my career, following the guidelines but doing as the boss tells us. Yes, indeed I was told that the workshop safety starts from the floor up, broken glass is one issue, and you could possibly step onto a small rod of glass and topple over if it were not seen, so my first job was to check the floor of the workshop that had about twelve glassblowers working away and of course ignoring my existence. I was pointed to the famous gatherer of floor glass... The broom, not just any broom, this broom had history, this broom had been in the glass business for at least 35 years before I was thought of, this broom had seen glassblowers come and go and had its own name. It was called Harry and often there would be a shout out for Harry. I was handed the broom like I was being given a royal sceptre and given the instruction to only pull toward me and not push the broom away from me. It was explained that Harry was to gather up the mess on the floor and not push it away and not push dust into the air. I took all this in and believed everything I was being told. He had his own little guard room, open type with no door and two hooks for the brush end to stand upright almost like a soldier would stand in with an instruction notice on the inside. Harry's Rules... Pull do not push, I'm not a garden broom, clean after use.

It was only a few weeks later that I found out, after hearing many shouts for Harry to be deployed, that Harry over the many years of service has had eight heads and twelve handles and that the term "Harry" was a way for the glassblowers to chase my tail. I must say, Harry did a very good job and that it was down to everyone that was supposed to call on his services and the new apprentice realized it was done all in jest. Harry finally retired when that company closed and was taken home by one of the workers and had his guard room placed in a garage. We have a new Harry now in my own business, and he is not as revered as the old Harry, but I still pull and not push him.

Regarding issues related to the H&S rules, the list can be endless when you consider the types of lathes, grinding equipment, burners, gas lines and most obviously the people that use them. I was given the daunting task of using a Skivener machine that ground blank glass keys. This machine had two rather large grinding wheels that would spin in opposite directions very close to each other, the idea would be to feed the glass blank key in between the wheels, pull on a handle and the wheels would grind the glass taper onto the key which was then to be finished on the grinding bench. I was warned about the wheels snagging the glass and the key firing out toward you... what I was not told was that this was a sort of game that was played, a key could be placed so it would deliberately snag and be forcefully regurgitated into many pieces requiring the help of Harry. That was not the real game, it was to see who could get the key as far away from the machine to score points. Yes, I know, it was a dangerous game having broken glass keys flying across a busy workshop. It was not the machine that caused the H&S issue, it was the operator.

Testing gas lines for leaks, I was informed that you don't wait to smell a gas leak, you test the line at least once a month. A soapy solution would be made up and applied to the joints using a small bush, if bubbles appear, close the line down and inspect the joint and replace if needed. That is until the Managing Director, a very accomplished glassblower in his heyday, informed us not to waste our time, he was informed that there was a possible leak near the main gas feed line, we were told to be on our way and he will have a look himself. As we walked away, we could hear a loud thump noise, we stopped, turned around and saw the M.D. walking away in a daze, he shut the gas mains off but not before lighting a piece of cardboard and knocking the flame out and putting the smouldering card to where we thought the leak was. He came out and said, "Yep, I've found the leak". That took a week to repair the gas line. There does seem to be a link here, one can measure intelligence, but one cannot determine levels of stupidity.

Stealth & Hasty is the art of being inside the rules of engagement, for example, one senior manager would often put cardboard boxes near the fire escape that had stairs for offices. Now this was a closed off area, protected by a fire door from our workshop into the stairwell. His train of thought was that if it is not allowed to be stored there, someone else would take it away and the situation would be resolved.

This very senior manager wanted his office redecorated and we were told to move everything out of his office on the Friday for the decorators to come in over the weekend. We asked where we should put all his "stuff", his reply was I don't care just empty the office. Que malicious compliance, we moved everything out as requested through the fire door and stored under the fire escape stairs. Monday morning, he was very pleased with his freshly refurbished office and asked me to go and find someone to help put back his desk and equipment. Well as if we couldn't have guessed it, it had all gone, once he calmed down after being told where it had been stored over the weekend, he had to track down the "janitor" and explain that it was him that kept using the space under the stairs. As luck had it for me but not for him, the waste compactor enjoyed a breakfast of desk, chairs and lamp. The stairwell was never used for storage again.

When is an accident at work an accident!

We all know that there are times when one thinks that the "rules" seem to look as if they are over the top, but there are times when a person must realize those rules were put in place for a reason by a person that must know just a little more on the subject matter than the person that has been assigned the job.

One such job, in a laboratory setting next to the glass department, was to extract nitric acid from mercury manometers that had been tested prior to packing. The task in hand involved removing the acid into a large glass vessel using a protected vacuum line, then changing the vac line on the manometer to draw distilled water through the glass into another large glass container, this would be done as a production line, remove acid, draw and remove water and the last process was to then suck acetone through the manometer and flush that through in a third container. All well and good, this process had been used for many years and the residue from the containers would be properly disposed of. This was a tiresome job that I had to do when the lovely lady lab assistant decided to go on maternity leave. I was fully aware of the implications and I was fully aware of the safety issues surrounding keeping the very expensive glassware in tip-top condition. Lovely lady then after a few months decided that she would not be returning back to work. I was called into the office to be informed that I would have to train a replacement and one would be found soon so that I could return to my glassblowing training, I did request that did not include calling on Harry, I was assured that was not going to happen again.

The job was advertised internally and an elderly guy, ex-navy, applied for the position, he was more interested in the pay than the type of work and was given the position. I was pleased that

I could break out from the Quality Control center once I handed the reins over.

The first day induction went well; he was very senior in years to me and was a retired petty officer in the Navy. Trying to explain the process was difficult enough, he would question every aspect of what seemed to be an easy process, of why this had to be done. However, after showing him a day's work and many visits from the laboratory supervisor and pointing out the flow chart that had to be followed, as long as that was in keeping, everything should be fine. Day two, he came to me and he showed me a homemade flow chart that he had made to speed up the process, I liked the idea that he was considering things like that, but I explained to him the limitations and the dangers in his plans and that it cannot and must not be used and the wall chart on the wall must be adhered to. I watched him for about an hour and he was very compliant, the lab super came in, very happy with things and signed me off from the laboratory and I finally broke free from the QA lab and I could return to the hot shop.

After lunch I heard my name called over the factory speakers to return to QA. On arrival, the Lab super asked if I could go into the cleaning station to sort out the new guy, I thought perhaps he had mixed up the different types of manometers that were going through the cleaning process, I donned my mask entered the room to see a dark brown mist coming from the sink area, Navy guy decided that the bottles were getting full and tipped the acid, water and acetone down the sink... That was a big no no, fortunately the sinks were emptied into a safety sump so nothing could enter the main drains. Gallons of water were used to flush the system and I had to sound the alarm to remove people from the area. Mistake number one was forgiven, as the lab super did not inform him not to do this, that made its way onto the instruction list. Navy guy apologized and after an afternoon of cleaning out the systems, things were ready for the next day.

What happened the next day finally put the nail into his position working in the QA lab, he decided he would use his method which resulted in him draining under vacuum all the acid out of the manometer and then immediately sucking acetone through the glass into the same container as the acid... What happened next was a sight to behold, The room immediately filled with thick dark brown smoke, the containers then started to become hot, I got called over to the QA Lab once again, the auto alarms triggered, people were leaving and the Navy guy was still carrying on, unaware of his surroundings, he was drunk from the night before and though he was doing a wonderful job of speeding the process up, he was pulled out of the laboratory and the fire brigades specialist chemical unit was called.

He has no idea about the danger of chemical reactions especially with a nitric acid and acetone mixture, when told it could has exploded in his face and he did not seem to give a hoot. The fallout from that incident resulted in the laboratory closure for ten days and a complete strip down and rebuild. This was because we were not sure if the mercury residue had been disposed of correctly prior to the acid wash. The company had incurred a very high fine from the local authority. Lovely lady, who was enjoying her maternity leave, was asked to return with the benefit of a large pay increase and things got back on track. Navy guy tried to sue the company for dismissal and was kicked out of court, he never worked again as it made the local newspapers and he basically tarnished his name and nobody would employ him.

Is it so when someone is not trained in the proper use of chemicals?

Or is it again, stealth & hasty!

There seems to be a theme running here, it could be that people are the most dangerous item in any surroundings. It's almost as if the workplace is the safest place to be as long as there are no people to interfere.

Everyone loves pranks, and as long as nobody gets hurt, it always seems to be harmless, like attaching a string to a sliding door and pulling the door open allowing a breeze to blow out a fellow glassblowers lamp, who then gets up to pull the door closed and then, magically the door opens.

Perhaps slowly make pieces of small glass items in production disappear, so when they have a count up they have to make more of the items, after they place them into the kiln, the missing pieces are added to those already there, confusion arises as they have made more than they thought. Nothing too serious and definitely harmless.

Sometimes the most innocent of things can have an outcome that nobody would expect.

The time a small football had been kicked through an open window from some members of staff having a kick about in the yard; "Wow ! that was lucky" the ball bounced around and knocked some glass over on a bench, it just happened to be the guy that kicked the balls bench, relieved that nothing was broken on his bench he proceeded to throw the ball toward the open window into the yard as the boss shouted what are you doing, distracted by hearing the boss, the ball hit the wall and bounced around the workshop and proceeded in smashing lots of work on others' benches, yet again the stealth and hasty rule kicked in.

We can see it in our daily lives, we drive on the correct side of the road, we make sure that our families are keeping well, we don't let children play with knives, but I am sure that we really don't take much notice of Health and Safety because it becomes natural. However, there are some among us that do know the rules, but like to engage Stealth and Hasty just to spice things up in their lives, often without thinking.

In my household we all know the simple rules in life. . .

Fun

Fun with consequences

Consequences
Health and Safety is paramount, Stealth and Hasty has consequences!

Chapter 6

Construction of a Ferrous-free Vacuum Cart for a 400 MHz NMR

Gary Coyne

California State University, Los Angeles, CA, USA

Citation

A chemistry experiment at California State University, Los Angeles, required a vacuum system to be hooked up to an NMR. This presented a variety of challenges all centered around the fact that you cannot have ferrous metals near an NMR machine: both because the ferrous metals can affect the readings and the extremely high-powered magnets in the NMR can violently attract any ferrous metals too close in proximity to the NMR causing them to (at a minimum) stick on the sides of the machine, or worse causing extensive damage. The challenge was to figure out a structure that could bring a vacuum to the NMR while keeping all ferrous metals a safe distance away.

Background

I was contacted by a chemistry graduate student at California State University, Los Angeles, who desired to study xenon gas combined with a chromatographic material in an NMR (Nuclear Magnetic Resonance) environment. There are several issues with this goal: Oxygen will quench an NMR signal and destroy the sample; xenon maintains a hyperpolarized state for a relatively short time; and in a hyperpolarized (excited state), xenon is extremely oxygen-sensitive and is quickly "de-excited" by contact with oxygen. Lastly, the experiment requires multiple samples to be run and a vacuum line connected to the NMR would facilitate that objective.

Thus, for all three reasons the experiment required the use of a live, dynamic vacuum system connected to an NMR: First and foremost was to keep the samples in an oxygen-free environment. Secondly, using gas transfer, it will be possible to supply the sample into the NMR as it is prepared. Lastly, such a setup allows multiple repetitions of the test with no change in setup.

NMR achieves its analytical abilities by affecting a small sample of molecules. Specifically, these samples should be presented as isotopes (atoms with extra neutrons) such as ^{2}H, ^{14}N, ^{13}C, and for this experiment ^{129}Xe and ^{131}Xe all have what is called "spin."

These isotopes resonate energy at specific frequencies when placed in a strong magnetic field. The molecules tend to align themselves in either of two directions: with or against the direction of the field. When a second oscillating magnetic field is directed at right angles to the magnetic field, resonance occurs and the spin flips. When the alignment flips, the molecules absorb and then release energy. Each isotope has a very specific "fingerprint" of released energy and by measuring the released energy it is possible to identify the molecules, their purity, and under some situations, their structure.

The greater the magnetic influence, the more information can be gleaned from the material yielding greater structural information of the molecule. As seen in the image of a sucrose molecule (Figure 1), a small region with no definition using a 90 MHz NMR has clear, easily identifiable imaging when using a 600 MHz NMR.

Thus, under the dictum "bigger is better," the larger the NMR (with a greater magnetic force), the better the resolution and the greater the information that one can receive during analysis. At this time, CSULA had three operational NMRs and a small, 200 MHz unit that is not in service. The units in service included a 300 Mhz, 400 Mhz, and lastly, a 600 Mhz. These can be seen in Figure 2. The NMR specialist, Ali Jabalameli, is standing next to each machine to help provide a size comparison.

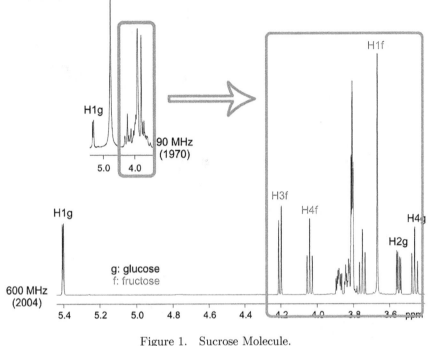

Figure 1.   Sucrose Molecule.

The Challenge

The challenge, however, is all ferrous metals must be kept at a distance of seven to eight feet to a point where the magnetic field is sufficiently weak as to have no effect. There are two main reasons for this and the obvious one is the effects of NMR on ferrous materials. For example, NMR specialist must remove any credit cards from their person, or it is a given that the magnetic strip on the cards will be wiped clean. Similarly, all watches, phones and other electronic devices cannot be brought close to NMR machines or they are likely to cease functioning.

In addition, larger ferrous objects must be kept at a distance from the NMR magnets for fear of being drawn to the magnets. For a side diversion, I suggest you do a Google-search for "chair stuck to an MRI" (Magnetic Resonance Imaging — used in medical examinations) for an entertaining and educational look at the

Gary Coyne

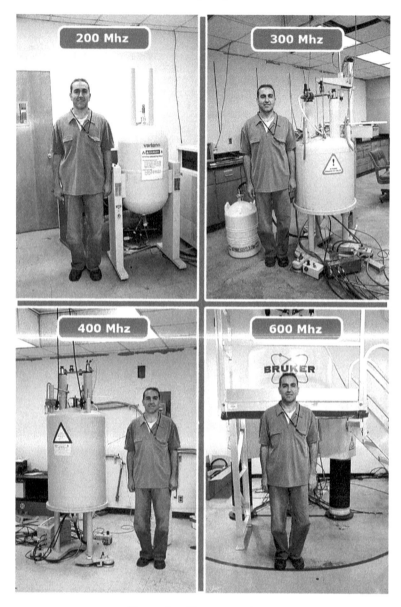

Figure 2.   Size comparison.

consequences. The magnetic field of an MRI is not too dissimilar to an NMR. Talk to any NMR specialist and you are likely to hear a bevy of "tales of horror" of wrenches, screws, chairs and a variety of metal objects stuck to an NMR.

The other reason why ferrous metals must be kept at a distance from an NMR machine is that they can affect the readings taken by an NMR. Thus, to keep the data reliable, ferrous materials must be kept at a distance from the NMR.

It was determined that the 400 MHz NMR was the most convenient for this series of experiments by reasons of practicality and accessibility. The height of the 600 Mhz was substantially taller and presented a host of complications. In addition, this was a new machine and everyone in the department wanted access making it harder to book time on this machine. That meant it was much easier to book time and access on the 400 Mzh. The challenge then was to figure out how to connect a vacuum system to the NMR and have the main components of the vacuum system sufficiently far away as to not cause any of the problems mentioned above.

One early approach considered was to attach a glass tube from the ceiling to connect to a small vacuum system already installed in the room. This approach can be seen in Figure 3. In this image you can see the path that would have been required by following the arrows.

There were multiple problems with this approach:

- There was nothing solid to connect to on the ceiling, since this was a dropped ceiling.
- The glass line would have crossed several recessed lights and if the lights needed replacement, the glass line was not likely to have survived.
- There would have been a permanent glass line hanging from the ceiling that would constantly and always be in the way of any other work on this NMR.

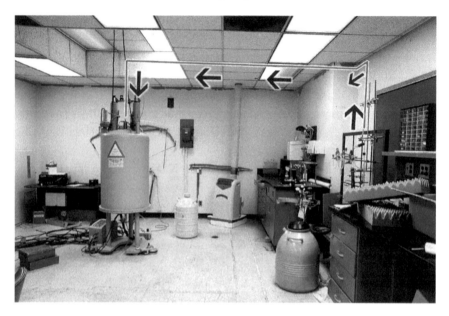

Figure 3.   Vacuum system path.

- The vacuum system in this room could not satisfy the throughput required by this new plan. We determined that tubing of at least 28–30 mm diameter would be needed to maintain the best minimum transport of gases.
- All of the right-angled turns would have degraded the quality of the vacuum throughput.

It was finally determined that a moveable, self-contained vacuum cart was necessary. The cart could either be built out of aluminum, stainless steel or wood. As woodworking is my hobby, the choice was obvious.

Both "needs" and "unknowns" governed the cart's final structure. The body was built in two parts: a large lower section (see Figure 4) that would contain the glass vacuum components, wheels and controls and a smaller top section that brought the vacuum to the NMR. The top part has a long extension to support the vacuum line to the NMR. The bottom section also had a sub-bottom to receive cinder blocks to provide extra weight as needed.

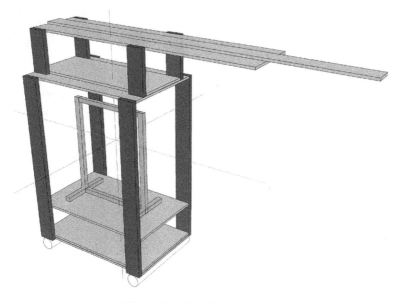

Figure 4. Initial cart design.

An additional reason for having a separate top part was because it was not practical to bring a standard metal tape measure in close proximity to the NMR. The limiting height was going to be the doorway height — we knew it was going to be close. By making the cart with a secondary top section, the cart was brought up to the NMR and one could easily measure the extra height required with a plastic ruler. Simply, it's much easier to get a very specific height for a foot or two than across eight feet.

On a side note, if the required height mandated that the cart be taller than a doorway, a removable joint connecting the glass line from the bottom to the top part would have been installed to allow for dismantling the cart to move from room to room. Fortunately, this was not necessary, as the final cart was able to clear a doorway by about 1/4th of an inch (about 6 mm). See Figure 5. [Note: the cart was stored two floors above where the NMR was located, but fortunately we had an equipment elevator that was sufficiently long and high enough to accommodate the cart.]

Figure 5. Cart clears the doorway.

Construction

There were two kinds of wood joints, or "joinery", used on this cart: dados and rabbets (see Figure 6). [Also seen in Figure 6 are the wheels used on the cart. The locking mechanism on these wheels not only prevents the wheels from rolling, but it also locks the wheels from swiveling. Despite the size and mass of the final cart, these wheels allowed very easy transport from the room where the cart was stored to the NMR room by a single person. Once locked in place, the cart could not be moved or jiggled from position. (Allowed as a compromise, the wheels became the only part of the entire cart where ferrous metals were used but were still located behind the safety limit lines.]

There were no bolts or screws or any other material to bring together the pieces of the cart beyond glue. In fact, when properly made, if a wood product is damaged, almost always it's the wood that fails before the glue joint fails. The joints' strength can be increased with properly selected joinery, in this case I used both dados and

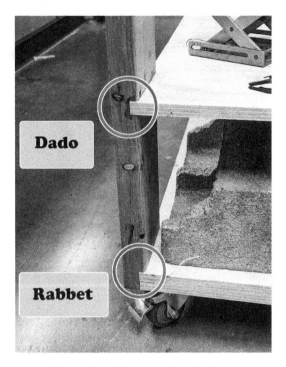

Figure 6. Carpentry terminology.

rabbets. Dados are structurally stronger joints than rabbets because of the extra wood surface that can be glued. To (mostly) assist with alignment, and provide a tad of extra strength, I used biscuits. Dowels would have provided greater strength than biscuits, but the extra work to get the dowels properly aligned was not justified for this purpose.

Biscuits, a biscuit cutter, and biscuits sitting in biscuit holes can be seen in Image #7. A fill-closeup image shows the blade extruded from the biscuit cutter.

Also shown in Figure 7 are the three common biscuit sizes (#0, #10 and #20) and a pair of dowels for comparison. The biscuit cutter is simply a small saw blade held horizontally and is pushed against the side of wood. How deep the blade cuts into the wood governs which size biscuit is used. By laying the biscuit cutter on a flat surface, it is easy to maintain a registered height for the biscuit and

Figure 7. Biscuits!

there is a tad of horizontal play to align the wood as needed before clamping until the glue has dried.

The Final Cart

The completed cart can be seen in Figure 8. It stands 6' 11–1/2" tall (2.121 meters), just short enough to get past doorways. The extension sticks out 7' 1/2" (2.146 meters) from the cart body with the overall length at 10' 1/2" (3.06 meters). The width, at 2' 6–1/4" (0.768 meters), is just narrow enough to get past doorways.

The metal lattice the vacuum system is attached to is made of aluminum as were all of the supports. The 2-finger clamps had small springs that had a very small magnetic attraction, so the springs were removed. Any ferrous metal parts were switched with stainless steel, aluminum or brass (with the exception of the wheels).

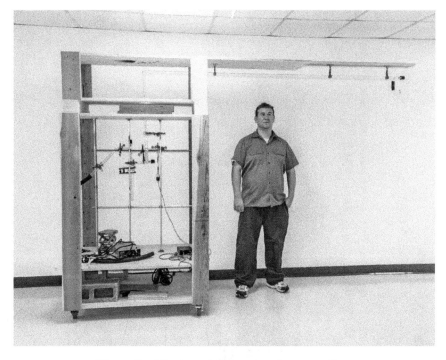

Figure 8. The completed cart and vacuum line.

Also seen in Figure 8, the glassware aspect of this vacuum line is nothing but the basics. There is a trap to keep pump vapors away from the experimental part of the line with stopcocks on either side of the trap. In addition, there is a thermocouple, and a stopcock on the end of the extension, and vents to release the vacuum for disassembly. In short, the glass aspect of this vacuum line is not particularly exciting.

Figure 9 shows the experimental probe that is slipped inside the NMR. The top part is essentially a modified single-port Schlenk line for evacuating the probe and backfilling with the activated xenon gas.

Figure 10 shows a close up of the end of the cart where the glassware is attached to the probe in place. Note that both the vacuum line and the probe each have a rubber hose between the bulk of each apparatus (seen within the two ellipses), and a set of

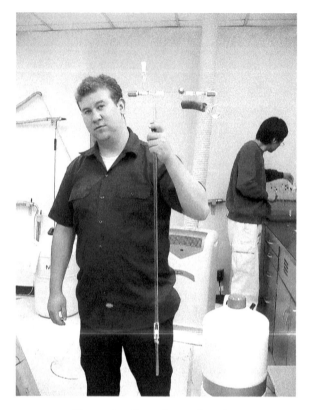

Figure 9.   Experimental probe.

greaseless ball-and-socket joints (at the arrow). This allowed for a reasonable amount of flexibility when attaching the cart to the probe, if you could get the two joints within a centimeter, you could easily join the two. Without this, it would have been essentially impossible to align the ball-and-socket joints up consistently each time. (On a side issue, note the nylon ropes attached to the NMR — suffice it to say we live in earthquake country.)

Figure 11 Shows the cart in position and attached to the probe apparatus. Note the two arrows on the bottom right of the image showing the "safe" and "very safe" lines on the ground (the further away the lines from the NMR, the safer the distance). The cart is several inches past the "very safe" region guaranteeing that nothing

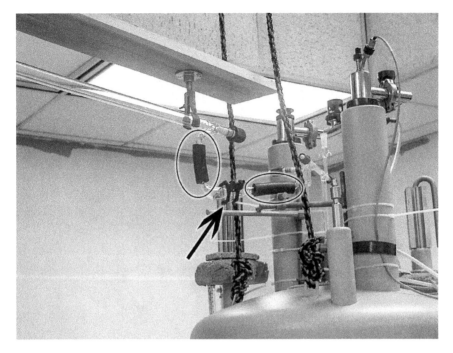

Figure 10.   Glassware with flexible connection.

on the cart should have any effect on the NMR readings and nothing on the cart is in a position to fling against the side of the NMR and stick. Also note the left side of the image showing the attachment of the mechanical pump to the cart's vacuum system. It was felt that placing the pump on a separate cart would reduce the pump's vibration than if the pump was placed directly on the cart. In addition, this also increased the distance as far as possible of the pump's ferrous metals from the NMR. It was hoped that the rubber hose connecting the two would attenuate any vibration. All evidence shows that this was the case: the end of the cart's extension arm displayed no vibration when the system was in operation.

Lastly, Figure 12 shows a better view of the pumping cart and its attachment to the vacuum cart. Note that the vacuum pump cart has the same locking wheels as the vacuum cart providing solid and stable locking wherever it is placed.

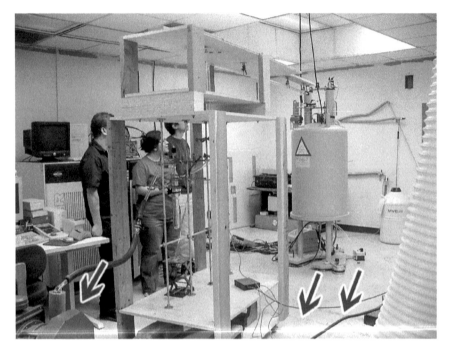

Figure 11. The cart in position.

Conclusion

The system did work as hoped with the one limitation that subsequent experiments took longer than hoped, but this was due to the time it took to evacuate the system after each run. Simply, the distance between actual experiment and the pumping system could not have been made shorter. Thus, the researchers just had to live with this limitation.

In conclusion, this chapter displays how one's hobby can directly aid and benefit one's employment. The glassblowing demands of this job were minimal and basic. In fact, knowledge of vacuum dynamics and requirements played more on the design than any glassblowing knowledge. The woodworking requirements were very basic and not overly challenging. This item was neither designed for beauty nor to follow any architectural trends. This project provided a classic

Figure 12.   The pumping cart.

example of how "just being a glassblower" would not have been sufficient for the job and how a complete glass shop utilizes any and all of a glassblower's skills to achieve the completed job.

Gary Coyne. California State University. Retired.

Chapter 7

Radiometer XXL, Nothing Is Impossible!

Peter Schweifel

Chairman of the German Scientific Glassblowing Society

It was an odd call to receive from Gallery Max Hetzler in Berlin. A work request for an exhibition by the British artist Darren Almond. I had been suggested as the final option. If I could not do it, the project would have to be dropped. A worldwide search for Glassblowers had only yielded negative answers so far, no one wanted or could make the desired piece. Now, with years of experience comes the wisdom of knowing what to make of such a compliment. Chances where it would not be a particularly difficult piece of work. More likely than not, many colleagues simply did not feel like getting involved in the project. "What's this about then?" I asked. They were on the search for someone who could build a radiometer. Convinced they were pulling my leg, I told them that these could be bought anywhere these days, for peanuts. "Well," they replied. "we need one that is 36 cm in diameter." Alright, that piqued my interest. I asked for some time to consider; you do not want to rush into a project like that!

A radiometer that size was not something I would build on my own. First, it is no fun doing it alone, and second, I was not sure I had the means to make this a reality without help. I had to find someone else who was mad enough to do this project, and who also happened to have the right skillset. Eckhard (Ecki) Martin came to mind.

Our phone call went something like this: "Ecki, do you know how to build a radiometer?" "Haven't tried it before, but I know the theory. And I have the tools to do it." — right, good start — "I am not talking about a small one, like you usually see, but something really big, 36 cm in diameter!" — silence. I waited. — "Well, if we don't try it, we'll never know if it can be done." That was decided then, we would attempt this project together.

For the uninitiated: Radiometers consist of a transparent glass bulb housing a bearing that carries four square wings connected by a star handle. The wings turn when the whole thing is exposed to light. They are usually a little larger than an old school light bulb. Not so for this project. The wing parts are made of mica plates — genuinely the size of plates, in our case. They would be $85 \times 85\,\text{mm}$. How would they behave? Would they still turn well at this size and weight, and if so, how fast? What sort of bearing would we need, and how do we mount the plates onto the bearing?

We started experimenting. Ecki still had parts for a radiometer from our late colleague Werner Herr, and we got hold of a large exsiccator to do some preliminary tests. These went well, albeit with smaller mica plates ($50 \times 50\,\text{mm}$). But what would happen to the wings under high heat while they are being encased in the bulb? In a balloon that size, you cannot keep an even heat. We started this second set of experiments "small" as well, with the same $50 \times 50\,\text{mm}$ mica plates. Tests with a 2-litre flask Failed Miserably! (Figure 1).

The aluminium star handle connecting the mica plates melted like butter. A new, different material was needed, titanium steel — anything heat resistant we could get our hands on. The star handles made of the different materials were connected mechanically with a small glass cap, which acted as the bearing. They worked! High temperatures were no longer an issue — one problem solved!

But larger mica plates posed a new problem. The plates at $85 \times 85\,\text{mm}$ weighted in at roughly $2\,\text{g}$ each. Though the production process for each was identical, there were weight differences of up to half a gram! The rotary wings kept tilting to one side inside the radiometer.

Figure 1. Failure!

What could we do? As 0.5 g was about a quarter of the plates, we could not just shave some material off to even out the weight distribution. That was not visually acceptable, this was for an art exhibition, after all! So, we ordered more plates and weighed them until we found four that were identical. Cost intensive, but goal oriented. However, this only solved part of the problem. The devil is in the details!

We had to think about where the weight was located within each mica plate. If the center of gravity of one plate was closer to the star handle than the center of gravity of the others, they were out of balance again. There was only one solution — keep trying until it works. So, we did.

After this, there was one more problem left to solve before the actual construction of the large radiometer. One side of the mica plates had to be covered in soot to create the force that would cause the wheel to turn. Covering something with soot is not a problem with a propane gas flame. As we all know, those flames produce soot like there's no tomorrow.

Soot has the unfortunate habit of burning at around 600°C. That is common knowledge and obvious to any glassblower. However, separating a flask with a volume of easily 20 litres into two parts to insert the wings, and then fusing the two pieces together again results in temperatures far exceeding 1000°C inside the flask. What were we to do? We knew that soot burns at around 600°C. In order to burn, it needs oxygen. Sublimation only happens far above our working temperatures, so to keep the soot from burning off, the oxygen had to go. We had to use inert gas and adjust the burners precisely, so that all added oxygen was burned up in the flame. These were problems we managed to solve (Figures 2 and 3).

Neither Ecki nor I had worked on such a large scale before — a flask of almost 36 cm in diameter was well outside of our experience. We did not even have the required machinery, the right kind of lathe, for this project. We needed additional help. It was time to involve a glassblower who knew how to work in these dimensions. Our immediate pick was Günter Diehm from Wertheim-Urphar. I had first met Günter at one of the workshops offered by the German Glassblowers' Association, Ecki had known him since adolescence.

Figure 2. Precise burner control.

Figure 3. The soot burnt off!

When we asked if he could build the part we needed, he replied "Piece of cake." Clearly the right man for the job! He was also willing to help us. "Come on down, we'll make a weekend of it." He said.

So, we headed to Günter Diehm's workshop in Urphar for the weekend, with our little prototypes stowed away in the car as models. Günter was well prepared, and the fusing went unexpectedly well despite the inert gas. We filled the object with inert gas again after fusing, and into the kiln it went.

On the following day, a Sunday morning, we met at Günter's workshop in high spirits. The mood shifted rapidly when we opened the kiln — the soot was gone; the inert gas had not protected it from burning off! Your guess is as good as ours. But for a glassblower, setbacks are a catapult toward progress. We opened the flask again, covered the plates with soot, fused the flask in inert gas, which successfully protected the soot at this stage, as it had before, and then what? — Do not put it in the kiln. Flasks can withstand unbelievable amounts of stress, so we decided to risk it.

In the following week, all that was left to do was to bring it to life. We took this thing of beauty to Ecki's workshop, where we could carry out our physical experiments with the exsiccator. The radiometer works better in a neon atmosphere, so we created a vacuum in the flask and filled it with neon. Created a vacuum again, filled it with neon again. Created a vacuum again and fused it. How high was the partial pressure? We do not really know, we just reduced the atmosphere until the wings moved well, then fused the thing shut and counted our blessings. The wings ran smoothly, and the movement was pleasantly slow, which fit the size of the object very well. Finished!

Or so we thought. We almost forgot that we had merely built a prototype. Prototypes often show errors that would have been difficult to predict. This was certainly true for ours.

Our final problem: The star handle had warped slightly during heating and, contrary to expectations, it had not returned to its exact original position. This meant that the mica plates did not remain perfectly upright and the wings tilted slightly to one side. The two radiometers that had been ordered could not have this issue, however invisible it was at first glance. This was art, after all!

The issue of the star handle in the middle of the rotary wings brought us back to our original battle with materials. At the end of a lot of trial and error, we ended up doing it completely differently, using a glass–metal connection and narrower rods than before. This worked faultlessly — nothing warped, and the connection could withstand the required temperatures (Figure 4).

With this last hurdle overcome, we later built two radiometers that looked perfect to us, one turning to the left, the other to the right. The first exhibition using our radiometers took place in Berlin as part of an installation titled "All Things Pass" by the UK artist Darren Almond. Both of us were invited to attend the exhibition, and, naturally, we accepted the invitation. It is important to enjoy the feeling of success that comes from accomplishing a difficult task.

Later, one of the radiometers became part of an installation in New York, the other in Barcelona (Figure 5).

Figure 4.   Star handle.

Figure 5.   Peter Schweifel, Gunther Diehm and Echard Martin.

The prototype went to the Geißler museum in Neuhaus am Rennweg as part of the Geißler year celebrations, where it gave joy to many visitors, despite small errors.

A few interesting questions remain.

The individual wings of a standard-sized radiometer have an area of about $2.5\,\mathrm{cm}^2$, and those radiometers turn very quickly. Our radiometer has an area of about $64\,\mathrm{cm}^2$ per mica plate, 26 times as large. It should turn even faster, but it does not! Why?

Contrary to many opinions, a radiometer does not turn due to the radiation pressure of light. It also does not turn due to higher gas pressure on the sooty side of the wings. So how does it work? That is an incredibly interesting question, and one for a different chapter.

Chapter 8

Toepler Pump, the Rise and Fall Of

Robert McLeod* and Paul Le Pinnet[†]

*Scottish Universities Environmental Research Centre, UK
[†]Fellow of the British Society of Scientific Glassblowers

For centuries glass has been central to the development of science and technology. Without scientific glassblowers to assist, many of the most important innovations in science and technology would not have happened. In the world of research nearly all the major breakthroughs involved glass somewhere in their development. Scientific glassblowers played a significant part in the development of Galileo's thermometer, Edison's light bulb, vacuum tubes for early radio, television and computers, fibre optics and lasers to name but a few of many. A specialist area where the skills and expertise of the scientific glassblower were put to good use was vacuum science and technology. The history and development of vacuum lines is a fascinating one. It can be traced back to ancient Greece. The Greek philosopher Democritus together with his teacher, Leucippus, may be considered as the inventors of the concept of vacuum and our modern view of physics is heavily influenced by their ideas. Their belief was that the empty space (in other words, in modern technology, a vacuum) existed between the atoms, which moved according to the general laws. Although early scientists made several advances in vacuum technology, there was still no clear definition of a vacuum. The word vacuum comes to us from the Latin word "vacuus" meaning empty or "vacare" to be empty. The greatest developments in vacuum technology happened in a fifty-year span

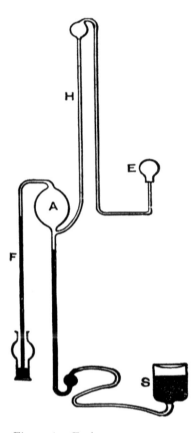

Figure 1.   Early mercury pump.

from 1850–1900 driven by the needs of scientific research and by the demands of the incandescent lamp industry. In 1865, the first pump using mercury was made (see Figure 1).

A falling droplet of mercury formed a piston which drove the air downwards.

Mercury was widely used in pumps for vacuum technology, one in particular being the Toepler pump.

The pump is named after August Toepler (Figure 2), a German lecturer of Chemistry and Physics. A glassblower, Herman Geissler, had invented a vacuum pump that was superior to the early piston pump that had been in use since the 17th century. The concept was

Figure 2.   August Toepler.

simple, the system to be evacuated would be connected to a bulb that could be filled with mercury and linked to a larger reservoir by means of a flexible rubber tube. Lowering the reservoir would draw gas out of the system and into the bulb. After closing a two-way tap to isolate the system, the reservoir could be raised again, compressing the trapped gas and pushing it out the system. Repeatedly lifting and lowering the reservoir led to extremely low pressures. It was slow and the timing of the turning of the taps had to be carefully coordinated with the mercury level. August Toepler simplified the pump by replacing the two-way tap with a T-junction using the mercury itself to act as a seal. On lowering the reservoir, the mercury would fall, exposing the T-junction, and drawing gas into the bulb. Raising the reservoir would trap a bulb full of gas and then compress it, allowing it to be extruded via a mercury non-return valve either to the open air, or more usefully into a gas burette (see Figure 3).

Scottish chemist William Ramsay, chasing the heavier noble gases, built vacuum lines with as many as eight Toepler pumps, but they were maddeningly boring to use and if one operation went wrong, the experiment had to be started again. Experiments such as Ramsay's drove the need for the Toepler pump to be improved

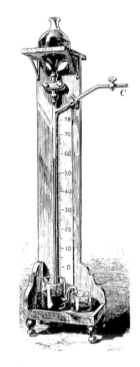

Figure 3.   Example of early mercury pump.

(see Figure 4). Some of the early glassware used by Ramsay which involved the skills of a scientific glassblower are shown in Figure 5.

From the early Toepler pump the design was improved so that the pump was more compact and more importantly automatic, which gave it many advantages over the previous designs. The mercury motion is now controlled by use of a switching relay which is activated by electrodes placed so that at the ultimate of its stroke the air leaked to the lower mercury reservoir is closed and an oil vacuum pump is started to evacuate it, and at the bottom of the stroke the vacuum pump is stopped and the air leak opened to allow the mercury to rise in the piston chamber. Pumping action is affected by two glass non-return valves with conically ground seats in series in the gas inlet and outlet tubes.

The wide acceptance of the new improved automatic Toepler pump meant that it proved useful in the quantitative transfer of

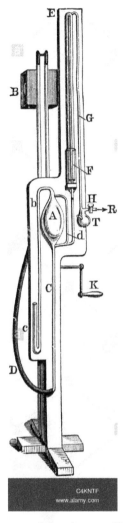

Figure 4. Toepler pump improved by August Toepler and used by William Ramsay.

gases and volatile liquids required for gas analysis. The new pump proved superior in the following respects:

(1) It is more compact and so requires less rack space for mounting.
(2) Better support for mercury reservoir ensures less danger of breakages.

Robert McLeod and Paul Le Pinnet

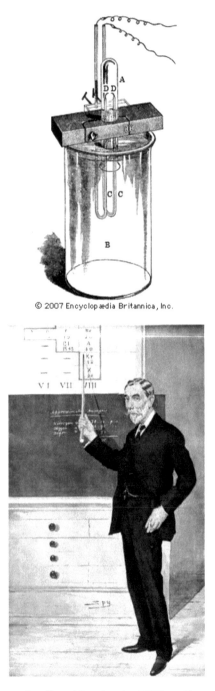

Figure 5.   Scottish scientist William Ramsay.

(3) It can be made with a much larger piston volume within a reasonable size with the shortest possible movement of mercury.
(4) The minimum movement of mercury allows the pump to be operated safely and smoothly at high speeds.
(5) The mercury is used efficiently since only a small amount of mercury in excess of the working piston is required.

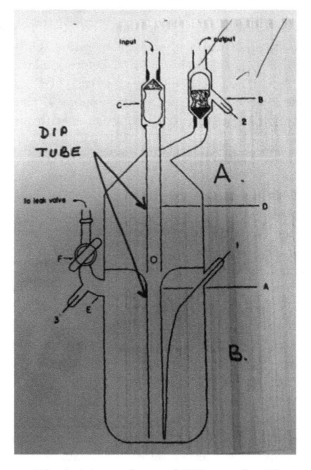

A pump with a piston volume of 800 cc and with an operating piston cycle of 35 seconds can evacuate a 2-litre system to 0.15% of original gas in approximately ten minutes. Pumps up to 200 cc piston volume size can be assembled by hand by a competent scientific glassblower. The larger pumps require use of a lathe for fabrication.

The pump can be made in two parts. (A) the upper piston chamber with dip tube and (B) the mercury reservoir. The two chambers are joined ensuring the reservoir is of a sufficient volume so that when in operation there is more than enough mercury to fill the piston chamber, the dip tubes and valves. The relevant arms are then sealed in place using a hand torch. Tungsten wires are sealed through the glass at strategic places (1–3 on Figure 2) so that the movement of mercury can be controlled. Care must be taken to ensure that the pump is flame-annealed after each operation. Once the pump is completed, it is annealed in the oven before joining to the line. The pump is mounted with the reservoir cast in Plaster of Paris or a suitable alternative to give added support. If the technology is available, the tungsten wire can be replaced with photosensitive cells which are put in place once the pump is joined to the line. They lessen the chance of stress fractures during construction of the pump. Care must be taken to position the line out of direct sunlight as the cells can be affected by the sun's rays.

At Scottish Universities Environmental Research Centre (SUERC), the main use of the Toepler pumps manufactured by the Glassblowing Workshop was in the collection and handling of small amounts of hydrogen gas extracted from geological (and other) samples for stable isotopic (deuterium/protium or D/H) determination (see Figure 6). Hydrogen, sometimes in the form of molecular water (H_2O), sometimes as hydroxyl (OH) and sometimes just elemental (H or H_2), is usually a minor constituent and the ratio of heavy hydrogen (deuterium ^2H, also denoted D, of mass 2, and naturally abundant at 156 parts per million, on average) to 'normal' hydrogen (protium of mass 1, ^1H) tells us much about the provenance or origin of the element in the specimen, and what natural processes (such as fluid-rock interaction) it has been subjected to. This often meant that the laboratory was examining rather exotic materials. Examples included meteorites (including some known to have originated on the planet Mars) and gemstones (hydrogen is present as an impurity in diamonds, and water is a constituent in the channels of beryl and emerald).

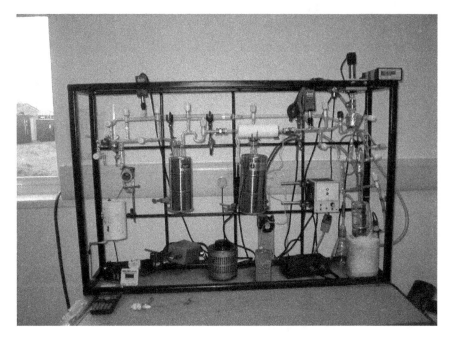

Figure 6.   Toepler pump in use at SUERC.

"NB as mercury is integral to the working of a Toepler Pump, operators must be aware of the potential dangers. In normal operation there is no danger of the operator being exposed to mercury vapors, which are toxic to the human nervous system and act as cumulative poisons, as the mercury is contained in a sealed unit. Danger occurs if there is a breakage. The room should be vacated and ventilated for at least 15 minutes. Specialist spill kits are available. The mercury spill kit contains a **dampening spray which when used will suppress the further release of mercury vapor**. Other components in the kit will then allow for the safe collection of mercury droplets."

My thanks to Professor Tony Fallick and Laboratory Manager Terry Donnely of SUERC for their assistance in putting this chapter together. Greatly appreciated.

Addendum: Word of the day "Toepler"

The Toepler pump has no moving parts and is an all glass construction, the only thing that moves is liquid Mercury which moves from an upper chamber to a lower and vice versa, all of which is conducted within a vacuum. The movement of the Mercury is used to transfer a specific volume of gas through the vacuum system.

On leaving school I was trained as an analytical chemist and therein lies a tale. The laboratory in which I was being trained was, to say the least, ancient and was always referred to as the old glass lab. The significance did not become apparent until much later, apparently there had been a glass works on the site since 1757. See the engraving dated 1869 which was originally called the Bank Quay glass works. After its closure in 1884, the area was dismantled except for the building to the left behind the three glass furnaces. This is where the laboratory was situated.

MERSEY FLINT GLASS WORKS,
ROBINSON, SON & SKINNER, *PROPRIETORS.* WARRINGTON.

It is interesting to note that Joseph Priestly had the glass domes for his pneumatic bath made at Bank Quay which is just half a mile from the Warrington Academy where he had conducted his experiments in "common air". To my mind this is an incredible series of coincidences!

I first encountered the automated Toepler pump which was used to calculate and check the surface area of a synthetic fluid catalyst via a series of manometers.

Watching a Mercury diffusion pump combined with a Toepler pump working within a vacuum system was far more interesting than the routine liquid/liquid analysis that I was involved with daily. Wondering how a Toepler and diffusion pump were made was the spur for me to train as a Scientific Glassblower.

It is a delight and a pleasure for me that Robert Mcleod Chairman of the British Society of Scientific Glassblowers has chosen without any prompting to write a chapter on the development of the Toepler pump.

For me this is a wonderful scenario, one which I could not have dreamed up. I thank him so much for his contribution.

Chapter 9

Calculation, Calibration and Hydrofluoric Acid Etching

Paul Le Pinnet

Fellow of the British Society of Scientific Glassblowers

The units used in modern calibration have their historical base in 1793. It was then that the Liter was first defined as a liquid measure, a convenient volume for commercial use. Toward the end of the 19th century, precise measurements were required in Chemistry, Physics and Engineering and in 1889 the "Standard Kilogram" was constructed, which was intended to be the same mass as 1 Liter of pure water at the temperature of its maximum density, which is 4 degrees centigrade. The Liter was then officially defined as the volume of 1 Kilogram of pure water at 4 degrees centigrade.

It was not until 1907 that a slight error was corrected. The 1889 "Standard Kilogram" was found to have a mass of 1000.028 cubic centimeters of pure water at 4 degrees Centigrade, it followed that the Liter was 1000.28 cubic centimeters! It was decided to leave the Kilogram as the "Standard" but to divide the Liter into 1000 equal parts and to call this division by a new name which was the Milliliter (ml). One Milliliter then equaled 1000.28 cubic centimeter and 1 cubic centimeter equaled 0.99972 milliliters. Therefore, from 1907 milliliters were used as the standard unit for liquid and volume measurement.

In 1964, the General Conference on Weights and Measures re-defined the Liter as equal to 1000 centimeters cubed (cm^3 or cc), this means that the Liter, by its new definition, is directly related to the Meter as a measurement of volume and no longer to the Kilogram. However, vessels calibrated in Milliliters (ml) are likely to be used for some time to come, and except for exactly accurate analytical work, it may be assumed that 1 milliliter (ml) equals 1 cubic centimeter (cm^3 or cc).

The Formula for Determining the Volume of a Cylinder

uses the ancient Greek symbol pi (π), which represents the ratio of the circumference of a circle to the diameter, which is 3.14, which means that the diameter of a circle (width) if multiplied by pi will be the same length as the circumference.

The formula for the volume of a cylinder is $\pi r^2 \times h$, where r is the radius, i.e. half the diameter, and r squared means that the radius is multiplied by its own value. h is the height of the cylinder.

This formula has been used in forming the nomograph (Figure 1), calculating volumes, diameters and lengths of a cylinder as follows: When given the cylinder's internal diameter and the volume desired, one must determine the length of the cylinder.

The solution is to place a straight edge on the cylinder's internal diameter on column 1 and the desired volume on column 2. This enables one to read the required length directly where the straight edge intersects column 3.

For example: A 10 mm internal diameter cylinder in order to contain a volume of 10 ml must be 127 millimeters in length.

Note: Any one of the three quantities can be determined given that two are known. Simply place a straight edge on the two known quantities and read the unknown where the straight edge intersects the third column.

The nomograph can be used for larger diameter tubing by multiplying each column by the power of 10.

Problem: What length of 50 mm internal diameter tubing is required to hold 1000 ml volume?

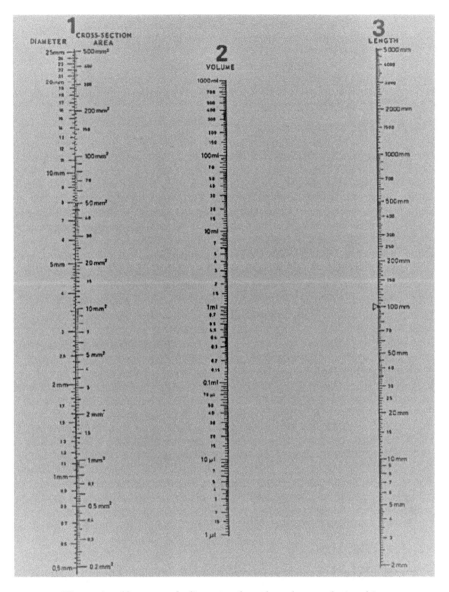

Figure 1. Nomograph diameter, length, volume relationship.

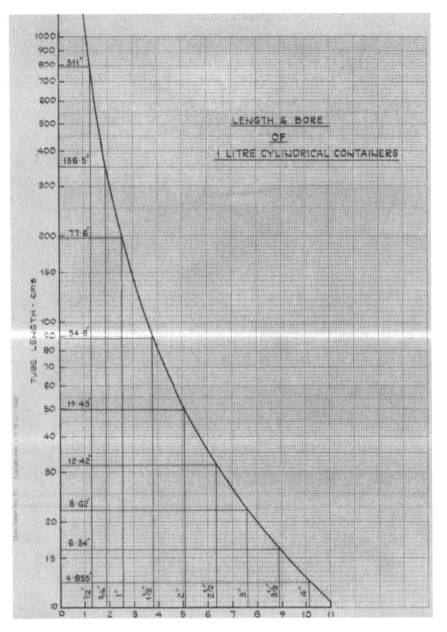

Figure 2.   Bore, length relationship/1 liter volume.

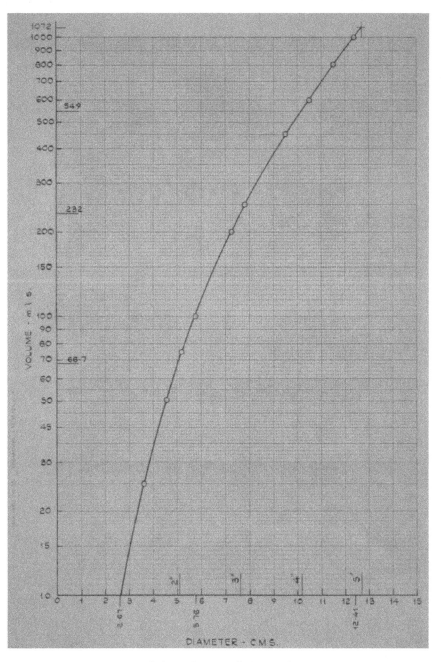

Figure 3.   Sphere diameter/volume relationship.

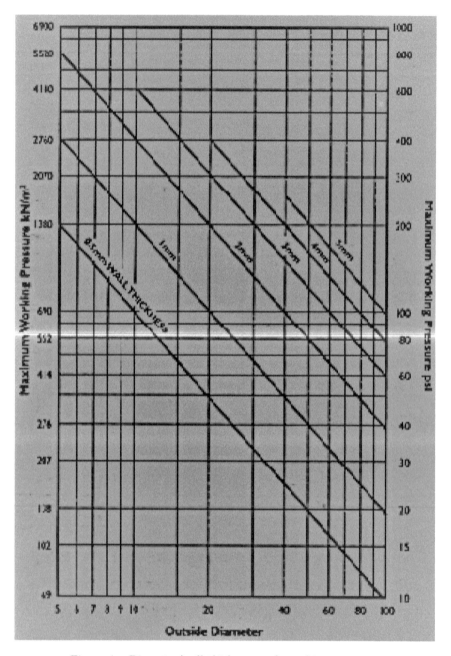

Figure 4.   Diameter/wall thickness, safe working pressure.

Solution: Place a straight edge on column 1 at 5 mm (1/10 of 50 ml) and on column 2 at 10 mm (1/100 of 1000 ml). The straight edge will intersect column 3 at 510 mm.

Therefore, the length of tube required for the required volume is 510 mm.

Hydrofluoric Acid Etching

The following method may be used where it may be difficult to accurately place on a decal or adhesive scale, such as on a tube that has bulbs along its calibrated length, or when there are two dissimilar diameters, or where a piece of apparatus is calibrated tap to tap.

When the tubing has been selected and the burette made up, the volume is confirmed against a known volume usually a class A burette, a mark is made with a fine pen using Indian ink.

A straight-sided burette usually only requires top and bottom marks, but when the burette is long, a mark every 10 ml is advised and if the calibrated length involves bulbs or tapers, then a series of calibrations need to be made.

Applying the resist for Hydrofluoric acid etching

There are many formulae for resists, including mixtures of beeswax, paraffin wax or ceresin. I have personally preferred beeswax only. The glass must be clean and free from fingerprints as these tend to raise the resist when brushed with acid, due to the salt present within the fingerprints. The glass is warmed, the hot wax applied with a squirrel haired brush, then allowed to drain and cool. If the glassware is cold when the hot wax is applied, the coating will be too thick. If the glass is too hot, the wax will tend to boil and drain away quickly, leaving a streaky coating of irregular thickness. The wax is then allowed to cool, and the cutting of the lines done the same day. If the wax is left longer, it tends to dry out and there is the possibility of the inscribed lines cracking along the edges.

The glassware is mounted upon a calibrating machine on which a variety of calibrated measures can be inserted depending upon the number of scribe marks that are to be made within a given distance.

The wax is scribed through with a spring-loaded point, which should be re-ground after each burette.

The length and frequency of the lines are governed by the British Standard BS846.

Etching

The etching is carried out in a fume cupboard, using a shallow polypropylene tray.

Thick long rubber gloves, an apron and a face shield must be worn.

It is essential that the fume cupboard is tested regularly as the Hydrofluoric acid fumes are deadly.

Next to the fume cupboard should be a shower.

A tube of Calcium Gluconate gel should always be near while working, with a second tube of the gel on the bench and a third tube at home, because the effect of dilute Hydrofluoric acid can be delayed in its appearance.

The etching process is carried out with 60% Hydrofluoric acid, which is wiped onto the glass with a soft squirrel haired brush for 6 minutes. The brush is soft and does not penetrate the wax. If the lines are not wiped constantly, blotching occurs.

The acid is neutralized with slaked lime, which will be described in greater detail later in this chapter under the heading "waste disposal".

The wax is removed by hot water or steam. The resultant groove is filled with powdered glass enamel and fired at annealing temperatures.

The British standard states that it must be permanent and suitable for the job.

Hydrofluoric acid

Anhydrous Hydrofluoric acid is a clear, colourless liquid that boils at 19.5 degrees centigrade. Because of its low boiling point and high vapor pressure, anhydrous Hydrofluoric acid must be stored in pressure containers. A 70% aqueous solution is a common form of Hydrofluoric acid, which is miscible with water in all proportions and forms an azeotrope (38.3%) that boils at 112 degrees centigrade.

Toxicity

Anhydrous or concentrated Hydrofluoric acid causes immediate and serious burns to any part of the body. Dilute and gaseous Hydrofluoric acids are also dangerous, although several hours may pass before the acid penetrates the skin sufficiently to cause redness or a burning sensation. Wearing clothing (including leather shoes and gloves) that has absorbed small amounts of Hydrofluoric acid can result in serious effects such as painful, slow healing skin ulcers.

Hazards from fire or explosion

Hydrofluoric acid is non-flammable. It is difficult to contain because it attacks glass, concrete and most metals — especially cast iron and alloys that contain silica. It also attacks organic materials such as leather, natural rubber and wood. Because aqueous Hydrofluoric acid can cause the formation of hydrogen in metallic containers and piping, which can present a fire and explosive hazard, potential sources of ignition such as sparks and flames should be excluded from areas using Hydrofluoric acid.

Handling procedures

It is crucial to ensure good ventilation by working only in a fume cupboard so that the working levels of 3 ppm are not exceeded. All contact of the vapor or the liquid with the eyes, skin, respiratory system or digestive system must be avoided by using protective equipment, which should be washed after each usage to remove any Hydrofluoric acid on it. Safety showers and eyewash fountains must be nearby. Everybody working with Hydrofluoric acid should have received prior instruction about its hazards and the proper protective measures and should know the procedure for treatment in the event of exposure (Reinhardt *et al.,* 1966).

Spills and leaks

The vapor of anhydrous Hydrofluoric acid and aqueous 70% Hydrofluoric acid produce visible fumes in contact with moist air. Spills of Hydrofluoric acid must be removed immediately to minimize the dangers of inhalation, body contact, corrosion of equipment and possible generation of hazardous gases. Spills must be contained and diluted with water. The resulting solution should be neutralized with lime before disposal.

Emergency treatment

If necessary, wear appropriate respirators and avoid personal contamination during rescue.

In view of the severity of burns and possible delay in painful awareness, all users of Hydrofluoric acid must have a tube of calcium gluconate gel and instructions for its use, both of which they are advised to keep at home to cope with the probable delayed effects. Additional tube of calcium gluconate should be kept in the workplace for first aid treatment.

In all cases, real or suspect, start immediately.

Skin contact

  (i) Wash the affected area with copious volume of water for one minute.
 (ii) Remove contaminated clothing — avoid further skin contact by wearing gloves.
(iii) Massage calcium gluconate to the affected area with repeated applications and continue this treatment for 15 minutes after the pain has stopped or the patient receives medical help.
 (iv) Irrespective of whether a burn appears to have occurred or not, emergency medical aid should be summoned. If a Hydrofluoric acid is suspected away from work, the gel should be applied and professional medical assistance sought immediately, even if self-medication has been applied at home, as further treatment may be required.

Inhalation

Remove the patient to a safe area and keep warm, treat any burns as above, encourage the suppression of coughing. In all cases summon medical assistance immediately. Apply artificial respiration if breath has stopped or shows signs of failing.

Apply external cardiac massage in the event of cardiac arrest. If breathing is labored or patient is cyanozed, give oxygen through a face mask.

Eye contact

Irrigate immediately thoroughly with water until medical assistance is obtained.

Waste disposal

Hydrofluoric acid is unlike most acids in that it is unstable and has an affinity with calcium, so when it is in contact with the skin, it will pass though the tissue of the body to link with the calcium of the bones.

Sodium Hydroxide and Hydrofluoric acid produce Sodium Fluoride plus water. There is a secondary reaction where Sodium Fluoride and calcium from whatever source produce calcium fluoride and sodium. Whereas neutralizing with slaked lime produces calcium fluoride plus water. Calcium Fluoride is a neutral and safe product which can be found naturally as fluorite or fluorspar.

To dispose of waste Hydrofluoric acid, add it slowly to a large volume of an agitated solution of slaked lime. This neutralized solution (pH between 6 and 9) is then added to excess running water.

Bibliography

Reinhardt, C.F. *et al.* (1966).

Chapter 10

The Life and Times of Otto Baumbach, Master Glassblower

Alan Gall

Archivist for the Institute of Science and Technology

Prologue

On Christmas day 1999 I received a gift from my wife. Not the usual socks and underwear, but a book about Ernest Rutherford, pioneer of nuclear science. I soon became absorbed by the account of this remarkable man and one chapter in particular caught my attention. When Rutherford arrived at the University of Manchester in 1907 he found that the physics department could call upon three individuals: an instrument maker, a tinsmith and a superbly skilled glassblower by the name of Otto Baumbach. I decided to find out more about the "support staff" who were vital for Rutherford's research.

I now go back to my days as a Liverpool schoolboy in the early 1960s. My Father had recently started a new job as a sales representative for a Manchester company. This enabled me to make occasional visits to the City. I took the opportunity to further my growing interest in chemistry by calling at various suppliers of laboratory apparatus. One firm that I particularly enjoyed visiting was The Scientific Glassblowing Company, then on Upper Brook Street. The proprietor, Harry Stuart, I found to be very obliging. A man running a business usually has more important things to do than pander to a schoolboy with only pocket money to spend. However, he always seemed to find time and if a relatively straightforward piece of glassware was requested, he would make it on the spot. What I did not know then, and probably something of small interest at the time, emerged some forty years later. Harry Stuart's predecessor in the business had been based at the University of Manchester and the origins of The Scientific Glassblowing Company could be traced back to Otto Baumbach's appointment as glassblower to the University.

Jumping in time, it is now a cold January evening in 2001 and I have just driven a short distance to Middleton, a suburb of Manchester. A stroke of luck with an internet search has led to the discovery that a person of great interest lives here on a modest housing estate and I have an appointment to conduct an interview. I ring the bell and an elderly gentleman with a slightly high-pitched voice opens the door. I am thrilled to meet Geoffrey Baumbach, son of Otto. After some pleasantries, the cassette recorder is plugged in and a microphone propped up between some books on the carpet. Geoffrey transports me back to the heady days when the University of Manchester held sway as one of the most important centres for nuclear research. All too soon, Geoffrey is beginning to tire and we decide to resume at a later date. But before parting company he hands me a bundle of papers and amongst the trove are letters: one from Robert Falcon Scott of Antarctic fame, another from a future president of Israel, and two from Rutherford. I am about to embark on a journey of discovery that will take me several years.

Introduction

The German state of Thuringia has a long association with the craft of glassblowing. As the German chemical industry prospered, particularly with the manufacture of dyestuffs from the early 1870s, so commercial scientific glassblowing followed suit. By the start of the twentieth century Thuringian glassblowers could be found in abundance, whereas in England their contemporaries were few.

The search for job opportunities brought Otto Baumbach to England where he would remain for the rest of his life. He became an Anglophile, although his admiration for British values was severely tested by internment in both world wars. A few years after his arrival at the University of Manchester in 1902 began a period of major advances in physics, spearheaded by the New Zealand-born experimental scientist Ernest Rutherford. The attraction of working under Rutherford brought students from far and wide, many of whom would go on to lead distinguished careers.

Internment during World War One cut short Otto's glassblowing activities, but led to his erstwhile employee, Fritz Hartwig, forming the Scientific Glassblowing Company. Otto lodged near to the house of John Charles Cowlishaw and married his daughter. Later, the creation of J. C. Cowlishaw Ltd for the production of thermometers circumvented the problems that beset "enemy aliens" trying to run businesses after returning from internment.

A significant development occurred around 1924 when the works manager from the Manchester firm Mather and Platt approached Otto with a proposal, which resulted in J. C. Cowlishaw expanding considerably to cope with the production of sprinkler bulbs for fire protection.

J. C. Cowlishaw Ltd is no more but the Scientific Glassblowing Company has continued in business to the present day.

Germany

Born 10 September 1882 in the village of Niederwillingen to Friedrich Carl Baumbach and Ida Olga Louise Agnes Schickel, Gottlob Otto Baumbach grew up during tumultuous times. What we understand as the country of Germany today did not always exist but formed by the unification of German-speaking states that had sometimes been in military conflict. Austria and some of these states were defeated by Prussia in 1866, which led to the formation of a North German Confederation. Further consolidation occurred after the Franco-Prussian war of 1870/71 with the establishment of the *Deutsche Reich*. United under Prussian control were the city states of Hamburg, Bremen and Lübeck, the kingdoms of Bavaria, Saxony and Württemberg, combined with thirteen duchies and five grand duchies. Prime architect behind the domination of Prussia in the new confederation was Otto Edward Leopold von Bismarck, the first chancellor of the Reich. Bismarck's epithet as "the iron chancellor" came from a speech he gave on 30 September 1862: "Not through speeches and majority decisions will the great questions of the day be decided ... but by iron and blood." Bismarck resigned in 1890 after clashes with Kaiser Wilhelm II over the Kaiser's desire for greater imperial expansion abroad. Thereafter, militarism increased and the first of five German navy laws (1898, 1900, 1906, 1908 and 1912) started the construction of battleships to rival that of the British fleet.

Niederwillingen is in the State of Thuringia, central Germany, along with the important glassmaking town of Jena. Then, as now, the village was in an agricultural area, with a mixture of forest and farmland, close to the towns of Arnstadt and Stadtilm. It lies on the Wipfra, a tributary that eventually discharges into the Elbe, one of the major rivers in central Europe.

Frederick, father of Otto, earned a living by hunting and selling wild game[1], a food speciality of Thuringia along with the famed Thuringian potato dumpling. When not hunting, he rode to the local tavern where he often over indulged and had to rely on his horse's sense of direction for the return home.

Otto Baumbach had two nephews, Otto and Rudolph Nauber, sons of his half sister Hetwig (or Hedwig). They later trained as glassblowers under Otto and provided a useful service for their uncle by procuring German glassware and manufacturing some parts while at a workshop in Ilmenau. On Hetwig's deathbed she asked Otto to look after her sons. Otto later recalled that in Britain there would have been a postmortem, but for some reason this was not carried out in Germany.

Training and First Employment

Ilmenau is a town to the south west of Niederwillingen, about 10–12 miles away depending on the route taken. Historically a mining town and later a centre for glass instrument production, it was the nearest

---

[1] Geoffrey Baumbach gave the example of roebuck (strictly speaking the term for a male of the European roe deer) but Otto's marriage certificate gives his father's profession as Clerk.

place with a facility for training glassblowers — the Thüringische Landesfachschule für Glasinstrumententechnik (Thuringia School for Glass Instrument Technology).[2] The trade school awarded Otto with a Masters Certificate in glassblowing and in 1901 he embarked on his career.

An interesting report about developments in fused quartz processing appears in a publication by the United States Geological Survey, 1905.

Three chemists in Germany — Herceus, Siebert and Kühn — have succeeded in blowing flasks of ordinary laboratory sizes from fused quartz. The mineral is melted in crucibles of iridium or iridium-ruthenium by the oxyhydrogen flame in a furnace of lime or magnesia. The difficulty in previous attempts has been that the quartz glass produced has been full of bubbles. But these escape if the quartz is kept in fusion for some time, and this can be done in a crucible of iridium which will sustain a temperature of $2,200°C$ ..."[3]

This indicates quite a high degree of competence on the part of the German chemists and if the two last named are Carl Siebert and Albert Kühn then the formation of the firm Dr Siebert and Kükn in January 1901 offered a job opportunity for Otto with prospects of advanced training. However, Otto did not stay long and according to Geoffrey Baumbach next moved to Leiden University. Figure 1 shows a receipt with the wording "Ontvangen van Prof. Dr. H. Kamerlingh Onnes de Somma van zestig guilders voor besoldering van 13 October te(h?) 13 November 1910" (Received from Prof. Dr. H. Kamerlingh Onnes the sum of sixty guilders for remuneration from 13 October to 13 November. Although the year appears to be 1910 it could also be a sloppy 1900, but neither date quite agrees with Otto's supposed progression. The year 1901 would have fitted in a lot better as it was then that Professor Onnes founded a school for instrument makers and glassblowers[4], a suitable institution to have employed Otto as a

---

[2] Letter from Geoffrey Baumbach to Thaddeus J. Trenn at the Max Planck Institute 23 October 1972.
[3] David Day (1906) p. 1346.
[4] K. Mendelssohn (1966) p. 75.

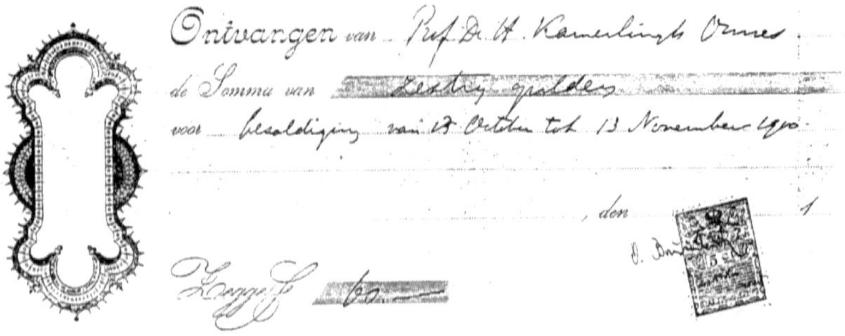
Ontvangen van Prof. Dr. H. Kamerlingh Onnes
de Somma van zestig gulden
voor bezoldiging van 18 October tot 13 November 1910
, den 1
Zegge f 60.—

Figure 1. A receipt for services rendered by Otto Baumbach to Professor Kamerlingh Onnes.

demonstrator of glassblowing techniques before his 1902 appointment at Manchester.

Winner of the 1913 Nobel Prize for Physics, Heike Kamerlingh Onnes (1853–1926) is famous for his low temperature research. He discovered the phenomena of superconductivity and is the first person to have liquefied helium.[5] Sir James Dewar (of vacuum flask fame) attempted the same liquefaction but abandoned further work on hearing of Onnes's success at Leiden in 1908.

Glass and Thuringia went hand in hand but, for a period, the employment conditions were not good, with poor remuneration offered for a ten to twelve hour day in blistering heat. The company history of a glassworks founded in Haselbach, Thuringia, gives an indication of the low wages paid in the late 19th century:

A marble maker earns from 50 to 60 Marks in 14 days, tube drawer working on piecework between 70 and 80 Marks. Annual room rent ranges from 90 to 100 Marks, a half litre of beer costs 12 pfennigs, four pounds of bread 40 pfennigs. Both tube drawing and the making of glass marbles are specialities of the Thuringian glassblowers.[6]

[5] Liquid helium boils at a temperature of 4.2 degrees above absolute zero at atmospheric pressure.
[6] *From Handcrafted Glass to High Tech Fibre* (Wertheim: Schuller GmbH, c. 1996).

Conversion to modern sterling values is very approximate at best. Taking a pre-WWI figure of 20 marks ≈ £1[7] and using the National Archives currency converter[8] one mark ≈ £4 in terms of today's values. The top-earners received £160 per week and paid about 50p for their pint of beer; not so bad.

Not everyone in Thuringia pursued a career in glassblowing but one estimate mentioned by Geoffrey Baumbach is that there were 25,000 in the region. Quite a good reason for moving to a location with little competition. "Those at the bottom of the pile [in Thuringia] made Christmas tree ornaments, those at the top, the most intricate complicated thermometers."[9]

Scientific Glassblowing in Manchester

The now defunct Orme Scientific Ltd is perhaps the first dedicated scientific glassblowing company to have operated in Manchester. Letterheads from the firm give a foundation date of 1900 and the first local directory listing is in the 1904 edition of Slater's Directory of Manchester and Salford, compiled the year before. Orme's first location at 56 & 58 Granby Row puts the operation in close proximity to the Manchester Technical School on Sackville Street, previously the Mechanics' Institution and later the University of Manchester Institute of Science & Technology (UMIST). Alfred Charles Orme arrived in Manchester from his father's laboratory supply business of C. E. Muller, Orme & Co Ltd of London. The lack of glassblowing skills in Manchester must surely have been the motivation for this move.

Richard Wagner, another glassblower from the Thuringia region, is known to have been employed as a glassblower since at least 1891 and went on to establish a business as thermometer and pyrometer makers under the name Wagner Brothers by 1910. There are other

---

[7]http://marcuse.faculty.history.ucsb.edu/projects/currency.htm [Accessed 12/07/2020].
[8]https://www.nationalarchives.gov.uk/currency-converter/#currency-result [Accessed 12/07/2020].
[9] Geoffrey Baumbach (1994).

contenders for earlier commercial scientific glassblowing in the region, companies that moved into the supply of scientific and laboratory apparatus which originated as chemists shops and possibly used glassblowers before 1900. Examples are James Woolley and John Mottershead. The latter can be traced to Mottershead & Brown in 1820, a partnership trading as chemists & druggists on Market Street. Mottershead took on various partners over the years but trading as Mottershead & Co had branched out into supplying chemical and photographic apparatus by the late 1850s. In a series of successions, Mottershead became Frederick Jackson & Co Ltd, J. W. Towers & Co Ltd acquired Jackson and Towers merged with A. Gallenkamp & Co Ltd.

James Woolley also started with a chemists shop and had the distinction of being a personal friend of John Dalton, renowned for his atomic theory. From initially retailing chemical preparations, Woolley moved on to the manufacture of pharmaceuticals. After Woolley's death in 1858, and by 1872, a scientific apparatus and surgical instrument department had been set up but it is not known exactly when facilities to manufacture glassware were established. This side of the business was eventually discontinued to concentrate on chemical manufacture. In 1962 British Drug Houses made a successful bid for James Woolley, Sons & Co Ltd.[10]

At the University itself, the task of glassblowing fell to any of the students and staff who showed some aptitude in the craft. Prior to the arrival of Otto the physics department relied on Arthur Stanton, private research assistant to professor Arthur Schuster. Unfortunately Stanton suffered from poor health and addiction to morphia which increasingly disabled his ability to construct complex apparatus. His addiction intensified after a traumatic incident when he assisted in taking X-rays of a woman's head after she had been shot at close quarters by her husband.[11] Despite the best efforts of staff, Stanton did not improve and died in 1898.

---

[10] For a history up to 1946 see *Woolleys of Manchester* (1946).

[11] See Alan Gall, "The Hargreaves Hartley Case: An Episode in the Early History of X-Rays", *BSSG Journal*, October 2018.

Otto Arrives at Manchester

A useful summary of the German community in Manchester and the prevailing attitude to Germans before WWI is given by Jonathan Westaway.[12] German culture and methods of education were much admired by the middle-classes of Manchester. For England and Wales the German community represented the largest group of foreigners up to 1891, and second largest up to 1914. The number of German-born residents of Manchester in 1891 stood at 1,321. German connections at the University of Manchester were particularly strong. Head of physics, Arthur Schuster (Franz Arthur Friedrich Schuster) had been born in Frankfurt am Main. Vice-chancellor Sir Alfred Hopkinson was married to Evelyn Oldenburg, who had a Bavarian father. Just about anybody who had a PhD degree had studied for it in Germany. For example, chemistry professor H. B. Dixon gained his doctorate under Robert Bunsen. W. H. Perkin Jr, organic chemistry professor, received his PhD for work supervised by Adolf von Baeyer.[13] The atmosphere was most decidedly pro-German — until 1914.

So Otto could expect, and received, ready acceptance in his new role. First, accommodation needed to be found within easy reach of the University and he secured lodgings at the house of Mrs Ellen Collins, 104 Heywood Street (later Harpenden Street), Moss Side, a mile away as the crow flies. It so happened that in the third house along, on the Corner of Heywood Street and Great Western Street, lived the Cowlishaw family. It seems probable that this proximity gave Otto the opportunity to meet a certain young lady. On 8 December 1906, at the age of 24, Gottlob Otto Baumbach married Hannah Lilian Cowlishaw, daughter of John Charles Cowlishaw and Emily (née Ashworth). Six months later came the birth of Rudolf who survived barely a month. There followed Otto Karl (b. 16 May 1910) Olga Emily (b. 1912) and John Geoffrey Baumbach (22 November 1921). The name Otto Karl did not go down well with the general population when World War One started and was replaced with

---

[12] Jonathan Westaway (2009).
[13] The naturally occurring dye Indigo was first artificially synthesised by Adolf von Baeyer in 1878.

Charles. The Christian names of the youngest child were selected as typically English and he mostly used his middle name of Geoffrey.

Geoffrey Baumbach stated that his father started business at 10 Lime Grove in 1902. The date is very credible but it is more likely that a bench at the University constituted the first base of operations. His listing in the 1905 local directory reads "Baumbach, Otto (scientific glass blower to the Victoria University) Chemical Department, Victoria University of Manchester, Oxford Road" and no occupant is listed at 10 Lime Grove in the same directory. By 1906 the premises at Lime Grove had been provided by the University and Otto, after his marriage, had taken up residence with his father-in-law at 34 Great Western Street.

It is believed that Harold Baily Dixon FRS (1852–1930), Professor and Director of the Chemical Laboratory, appointed Otto as glassblower[14]. Dixon, an accomplished Alpine climber, specialised in combustion research. He is also noted for disputes with Ernest Rutherford. In one incident, Rutherford reported:

I have had some little trouble with friend Dixon. He saw Pring (one of the lecturers) and complained that his lectures were not up to standard ... He had forgotten to see me first in his usual way, so next day he had to metaphorically crawl in my office. I gave him a little to chew over in private for some time to come."[15]

It would be surprising if competition for Otto's time did not feature in the wrangling since both the chemistry and physics departments had need of specialised glassware, as did others in the science faculty. The usual arrangement for such services was that the University had first call. All three technical service providers, the glassblower, instrument maker and tinsmith, were allowed to take on outside work as long as research and teaching requirements were met. In return the University provided premises and guaranteed minimum earnings.

---

[14] Although Per F. Dahl (1997) states that Otto had been "imported from Germany by Schuster".

[15] David Wilson (2003) p. 229. From a letter written to Arthur Schuster in May 1908.

On Otto's arrival, The Victoria University consisted of three colleges: Owens College, Manchester, University College, Liverpool, and Yorkshire College, Leeds. Independence came in 1903 to create the University of Liverpool and the Victoria University of Manchester, the University of Leeds in 1904.

Living and Working in Manchester

Otto's lodgings at Moss Side and the Cowlishaw household stood in an area of dense population. A count of houses to the south of Moss Lane West indicates that around 1800 houses once existed on 30 separate streets in an area of just over one tenth of a square mile.[16] A far cry from the beautiful Thuringian forests. Pollution and poor weather conditions were also features that caused comment by workers at the University. Henry Moseley (of whom, more later) wrote to his mother on 30 November 1910.

Today the fog is so thick, that I shall probably get lost on my way to the College; it tastes acrid and tickles the throat. Yesterday the tram in which I came back lost itself badly, and I finally got out somewhere and groped for a side-street on which to find a street name which luckily I recognized. Monday the fog was thinner but more yellow.[17]

The Finnish scientist Lars Öholm spent 1913 working in Rutherford's laboratory. A letter sent to the Nobel Prize-winner Svente Arrhenius gave his impression of the city. In translation it reads:

As you see I am now in Manchester and attend the course in radioactive measurements. I have also got a picture of what an industrial town means. Everything is sooty and dirty. Perhaps you remember how it was when you visited here. Moreover, the weather

---

[16] A calculation made by the author for an article on the Moss Side Brewery in *Manchester Breweries of Times Gone By* (Salford: Neil Richardson, 1982).

[17] J. L. Heilbron (1974) p. 178.

has been quite bad. The laboratory is not very attractive either. Soot and smoke have left footprints everywhere.[18]

Otto Baumbach & Co prospered well enough for the Baumbachs and John C. Cowlishaw to relocate from Moss Side to Hale, in the Cheshire countryside, around 1910. The house stood on Alan Drive, one of nine detached properties on a short stretch of road, and was christened with the name "Waldruhe". The affluence extended to the services of a domestic servant and a Ford model T motorcar, but there were shattering events waiting to unfold. The Nauber brothers, Rudolf and Otto, who were being trained by their uncle, had to make do with lodging at the house of Mrs Collins. In 1910 they were still in their teens, Rudolf aged seventeen and Otto only thirteen.

Otto gave demonstrations of his art. At the annual soirée of the Manchester [students] Union: "In the Frankland laboratory Mr Otto Baumbach was blowing glass into pretty shapes and colours".

He also made ornaments for Rutherford's wife Mary, who wrote in appreciation:

Dear Mr Baumbach, I have just returned from holiday and find here the very beautiful menu holders that you have so kindly made for me. It was very good indeed of you to take so much trouble for me and I admire them greatly.[19]

On 1 March 1912, Arthur Schuster officially opened extensions to the Physical and Electrotechnical Laboratories. The occasion merited a pamphlet which described workshops in some detail.

Recently a special building has been erected adjoining the new Electrical Laboratory, in which the University glass-blower, Mr. Baumbach, and the University tinsmiths, Messrs. Stelfox, are housed. Special workshops of this character are essential for the efficient working of the scientific Departments of the University.

---

[18]Peter Holmberg (2016) p. 23.
[19]Letter from Mary Rutherford dated 16 August, but without the year. Original with the Baumbach family.

Figure 2.   Ernest Rutherford in the later Cambridge days (John Rowland, 1955).

Rutherford

The most comprehensive account of the life of Ernest Rutherford (Figure 2) is John Campbell's *Rutherford: Scientist Supreme* (1999) and is a highly recommended read. The earlier book by David Wilson (1983) is also a most useful source of information. Only a very brief sketch needs to be given here.

A New Zealander by birth (30 August 1871), Rutherford gained what was called an 1851 Exhibition Scholarship to study under J. J. Thomson (discoverer of the electron) at the Cavendish Laboratory, University of Cambridge, arriving in 1895. He could have made his mark in radio, later taken to commercial success by Marconi, but the discovery of radioactivity by Henri Becquerel in 1896 changed his focus. Rutherford then identified two different types of radiation — alpha and beta. His experimental work was judged to be of a sufficiently high standard to secure a professorship at McGill University in Canada. Attracted by an exceptionally well-equipped laboratory, and the chance of becoming a professor at the young age of 27, he left Cambridge in 1898. At McGill, and working with British chemist Frederick Soddy, Rutherford completed experiments that would earn him the Nobel Prize in Chemistry "For his investigation into the disintegration of the elements and the chemistry of radioactive substances".

Arthur Schuster, the Langworthy Professor of Physics at Manchester, had been following the scientific progress of Rutherford. Such was Schuster's esteem for Rutherford's capabilities that he decided on an early retirement if Rutherford could be persuaded to accept the Langworthy Professorship. A respectable salary and the desire to be closer to European scientific developments brought the Rutherfords (he had married Mary Newton in 1900) to England.

At the turn of the century, the Faculty of Science at Manchester could boast some illustrious names.

When Rutherford arrived, William H. Perkin, son of the discoverer of the first aniline dye, headed the chemistry department. Osborne Reynolds, of Reynolds number fame,[20] held the Research Professorship in Civil and Mechanical Engineering. Horace Lamb, a leading expert on hydrodynamics and an influential contributor to theoretical seismology, was a professor of mathematics and reputedly one of the finest applied mathematicians of his time. Others were just beginning their careers. A demonstrator in chemistry and a research worker called Chaim Weizmann would gain fame as the first president of Israel, as well as by making a significant contribution to the war effort in WWI. Schuster's assistant, Hans Geiger, would become a household name through his work on the famous radioactivity counter. James Chadwick, then a physics student, later proved the existence of the neutron. The years before World War One saw many other additions. Georg von Hevesy, a future Nobel Prize-winner for radioactive tracers, was awarded a research fellowship in 1910. The grandson of the evolutionist Charles Darwin, also called Charles Darwin, replaced Harry Bateman as Reader in Mathematical Physics, and in turn was replaced by quantum theory maestro Niels Bohr. Henry Moseley enjoyed a short but productive career, sometimes

---

[20]The Reynolds number is an indicator of when fluid flow in a pipe turns from laminar to turbulent flow — used in design calculations by chemical engineers and by civil engineers.

working with Kasimir Fajans (now remembered for Fajan's rules in chemistry), before an untimely death in the battle for Gallipoli, just two months short of his 28th birthday. Had he lived, his achievement in using X-ray spectra to place the elements in order of their atomic number would surely have earned him a Nobel Prize. Most of those named, and many more besides, were drawn to Manchester by the opportunity of working with Rutherford. He could be a hard task master but the academic rewards were great.

Rutherford took up his post at the University in May 1907 and impressed the technician William Alexander Kay[21] by bounding up the stairs, two at a time, to his office. Kay and Rutherford got on well together especially as Kay had a gift for understanding the workings of scientific apparatus and the ability to put things right when problems arose. It is clear that Rutherford also appreciated the capabilities of Otto. After Rutherford left Manchester for Cambridge, he had occasion to use Otto's services until persuaded to hire Felix Niedergesass as the official University of Cambridge glassblower.

Writing for the *Radio Times*, Professor T. H. Pear gave his personal view of Rutherford's character.

When he was not talking physics, and often he was not, for he was very human, he displayed a lovable simplicity of mind, loosing off on any unprepared subject under, or over, the sun. Like the output of those radio sets which take a little time to heat up, his first answer to an unexpected question often lacked warmth, but when the new idea dawned he burst into flames.[22]

Some Scientific Endeavours

An early example of Otto's contribution to experimental studies in the chemistry department is contained in a paper by Professor Dixon and E. C. Edgar, for work completed no later than 1905.

---

[21] William Kay's association with Rutherford did much to enhance his reputation. In acknowledgement of his services, a University building was named William Kay House and he received an honorary MSc.

[22] T. H. Pear, "University Life in Manchester", *Radio Times*, 25 May 1951.

The Belgian analytical chemist Jean Stas had made a determination of chlorine's atomic weight but used a sequence of steps that might have introduced cumulative errors. Dixon & Edgar investigated a one-step method, finding a slightly higher value than did Stas. Setting up the apparatus took the best part of two years and achieved an important requirement — that the glass taps be perfectly gas-tight (interestingly, these taps was lubricated with molten glacial acetic acid).

We are indebted to the skill of the University glass-blower, OTTO BAUMBACH, for the accurate grinding of these taps, and for the joints by which he succeeded in fusing hard Jena to soft glass.[23]

F. P. Burt & E. C. Edgar, senior lecturers in chemistry at Manchester, heaped on even more praise.

The junction between the thick platinum tube and the glass was an excellent piece of workmanship carried out by Baumbach, the University glass-blower; it was quite free from air-bubbles and absolutely gas-tight. For a year it has been heated and cooled, and no sign of a crack has ever appeared.[24]

Alpha radiation is helium without its two electrons, thus carrying two positive charges. Rutherford knew this but needed to establish the fact beyond doubt. To do this he enlisted the aid of Thomas Royds, a Manchester graduate with a first class honours physics degree and a recently awarded MSc. The results, completed in 1908, appeared in a paper sent to the *Philosophical Magazine* under the title "The Nature of the α Particle from Radioactive Substances". A diagram of the apparatus is shown in Figure 3 and the authors placed particular emphasis on the 15mm long tube labelled 'A'.

This fine tube, which was sealed on a larger capillary tube B, was sufficiently thin to allow the α particles from the emanation and its products to escape, but sufficiently strong to withstand atmospheric pressure. After some trials, Mr. Baumbach succeeded in blowing such fine tubes very uniform in thickness. The thickness of the wall of the

[23]Harold B. Dixon & E. C. Edgar (1906).
[24]F. P. Burt & E. C. Edgar (1916).

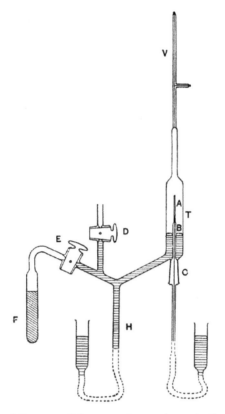

Figure 3. Drawing of the α particle apparatus as shown in the Rutherford-Royds paper of 1909.

tube employed in most of the experiments was less than 1/100 mm., and was equivalent in stopping power of the α particle to about 2 cms. of air.[25]

The actual apparatus as constructed by Otto is shown in Figure 4. Per F. Dahl (1997) has gone so far as to state that the experiment is sometimes referred to as the Rutherford-Royds-Baumbach experiment.[26]

---

[25] Ernest Rutherford & Thomas Royds (1909) pp. 282–283.
[26] Per F. Dahl (1997) p. 332. The author has found no other references specifically naming the Rutherford-Royds-Baumbach experiment. A request to Per F. Dahl for information on sources did not receive a response.

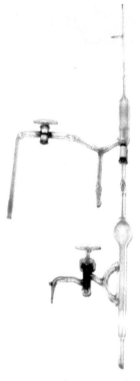

Figure 4. Apparatus for determining the nature of alpha particles. Constructed by Otto Baumbach in 1908. ©Cavendish Laboratory, University of Cambridge. Courtesy of the late Keith Papworth.

Radium produces the radioactive gas radon. In 1908 this gas had yet to be identified and went under the temporary name of "radium emanation". Alpha particles are less energetic than beta and gamma radiation and after passing through Otto's thin-walled tube were halted by the thicker glass of the section marked T on the drawing. After a period of time, the trapped particles reverted to atomic helium which could be detected spectroscopically. Further experiments discounted the possibility that traces of helium had leaked in from the surrounding air.

Moseley's work on high frequency spectra of elements required X-ray tubes. He acquired a number from the London manufacturer A. C. Cossor but came to rely on tubes made by Otto, which were considerably cheaper. The Solvay Institute funded Moseley's investigations and a report on expenses included "X ray tube from Baumbach £1.16.0"[27]

Other departments at Manchester needed the services of Otto. Charles Powell White of the pathology department wrote: "The glass regulating apparatus, which must be made of Jena glass with hollow plugs in the stopcocks, was made for me by Mr Otto Baumbach, 10 Lime Grove, Manchester." Before the advent of Pyrex and other British-made borosilicate glass, Jena was a popular choice. An advertisement in the 16 November 1911 issue of *Nature* lists the British stockists of Jena laboratory glass, including Otto Baumbach at Lime Grove (Figure 5).

It is very likely that a recommendation by Arthur Schuster landed Otto an order from Robert Falcon Scott, for 1200 sample tubes to taken on the ill-fated British Antarctic Expedition of 1910 (Figure 6).

A Tramload of Professors, 1913

On the occasion of the 100th anniversary of the University of Manchester an article with the above title appeared in the *Manchester Guardian*. W. M. Calder, author of the reminiscences in 1951and professor of Greek before he left for Edinburgh University in 1930, recalled his daily trip by tramcar. It gives a nice snapshot of the professors who would have been familiar figures to Otto. Although Rutherford owned a car, bought with his Nobel Prize money, it was convenient to use the transport that ran past his house. In the following I have added notes to selected quotes and indicated in brackets the period the over which the professorships were held.

"First into it — for he had invariably missed the one in front — was Herford, whose dreamy gaze gave no inkling of the critical acumen which cut through fustian like shears."

The journey to the University began at the Didsbury end of Palatine Road. Charles Harold Herford (1901–1921) was Professor of

---

[27]Kristen Frederick-Frost (2012) p. 73.

**The NEW
Jena Laboratory Glass**

Is on sale with the following firms—

London - - - - BAIRD & TATLOCK, Ltd., 14 Cross Street, Hatton Garden, E.C.
F. E. BECKER & CO. (W. & J. George, Ltd., Successors), 33 to 37 Hatton Wall, E.C.
A. GALLENKAMP & CO., Ltd., 19 & 21 Sun Street, Finsbury Square, E.C.
TOWNSON & MERCER, 34 Camomile Street, E.C.

Aberdeen - - - C. G. FRASER, 16 Hadden Street.
A. & J. SMITH, 23 and 25 St. Nicholas Street.

Birmingham - - F. E. BECKER & CO. (W. & J. George, Ltd., Successors). Great Charles Street (near Town Hall).
PHILIP HARRIS & CO., Ltd., Edmund Street.
STANDLEY BELCHER & MASON, Ltd., Church Street.

Bradford - - - HARRISON, PARKINSON & CO.,

Dublin - - - - PHILIP HARRIS & CO., Ltd., 179 Great Brunswick Street.
J. M. MAIBEN & CO., 31 Eden Quay.

Edinburgh - - - A. H. BAIRD, Lothian Street.
BAIRD & TATLOCK, Ltd., Teviot Place.
WILLIAM HUME, 14 Lothian Street.

Glasgow - - - BAIRD & TATLOCK, 45 Renfrew Street.
THOMSON, SKINNER & HAMILTON, 38 Sauchiehall Street.

Hull - - - - - E. B. ATKINSON & CO., 24 Dock Street.

Leeds - - - - REYNOLDS & BRANSON, Ltd., 14 Commercial Street.

Liverpool - - - BAIRD & TATLOCK, 164 Brownlow Hill.

Manchester - - BAIRD & TATLOCK, 25 High Street.
OTTO BAUMBACH, 10 Lime Grove, Oxford Road.
FREDK. JACKSON & CO., Ltd., 14 Cross Street.
JAMES WOOLLEY, SONS & CO., Ltd., Victoria Bridge.

Newcastle-on-Tyne BRADY & MARTIN, Ltd.

Sheffield - - - J. PRESTON, 105 Barkers Pool.

Stockton-on-Tees BAINBRIDGE & FINLAYSON, 141 High Street.

Widnes - - - - J. W. TOWERS & CO., Ltd.

Figure 5. Otto Baumbach is listed as a stockist of Jena glass in this advertisement from *Nature* 16 November 1911.

English Literature, living on Parkfield Road, Didsbury. He married German-born Marie Catherine Berge and named his son Siegfried.[28]

"Next, well up in years but blithe and debonair, would come Moulton, then completing his work on the great "Vocabulary" and turning to the affair with the Zoroastrians which was soon to lead to his tragic end."

The Rev James Hope Moulton combined his duties as a tutor at the Wesleyan College, Wilmslow Road with that of Greenwood Professor of Greek. He appears to have lived at the Wesleyan College

---

[28] Jonathan Westaway (2009) p. 599.

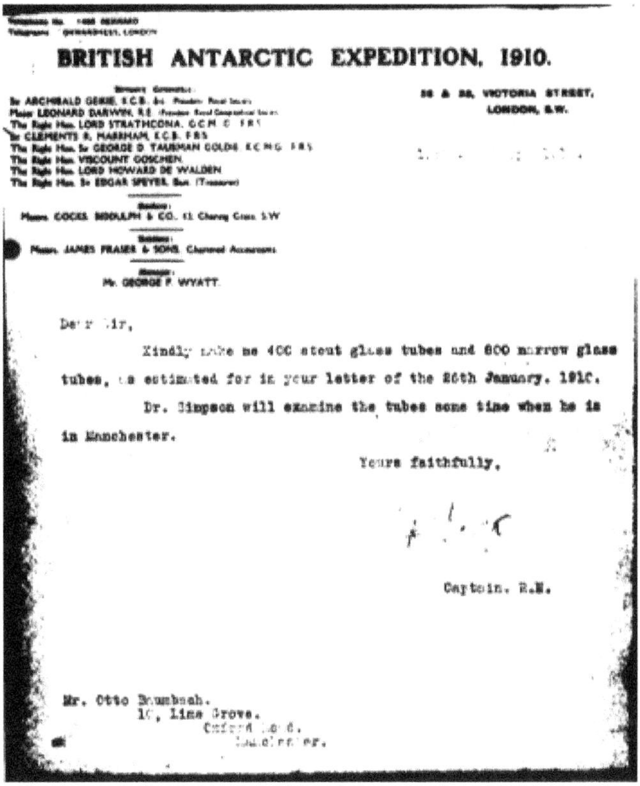

BRITISH ANTARCTIC EXPEDITION, 1910.

36 & 38, VICTORIA STREET,
LONDON, S.W.

Sir ARCHIBALD GEIKIE, K.C.B.
Major LEONARD DARWIN, R.E.
The Right Hon. LORD STRATHCONA, G.C.M.G. F.R.S.
Sir CLEMENTS R. MARKHAM, K.C.B. F.R.S.
The Right Hon. Sir GEORGE D. TAUBMAN GOLDIE, K.C.M.G. F.R.S.
The Right Hon. VISCOUNT GOSCHEN.
The Right Hon. LORD HOWARD DE WALDEN.
The Right Hon. Sir EDGAR SPEYER, Bart. (Treasurer)

Messrs. COCKS, BIDDULPH & CO., 43 Charing Cross, S.W.

Messrs. JAMES FRASER & SONS, Chartered Accountants.

Mr. GEORGE F. WYATT.

Dear Sir,

Kindly make me 400 stout glass tubes and 800 narrow glass tubes, as estimated for in your letter of the 26th January, 1910.

Dr. Simpson will examine the tubes some time when he is in Manchester.

Yours faithfully,

Captain, R.N.

Mr. Otto Baumbach.
10, Lime Grove.
Oxford Road.
Manchester.

Figure 6. An order for sample tubes signed by Captain Robert Falcon Scott. Courtesy of Philip Baumbach.

so might have walked 1.5 miles to the tram. The reference to his ‘tragic end’ concerns a trip to India during World War I when he decided to speak to a religious sect there called Parsees (believers in Zoroastrianism). On his return journey from a stopover in Egypt, the ship was torpedoed. It is said that he died of exhaustion while helping others.

“And with him Conway, dark, tall, lithe, eager, generous.”

Robert Seymour Conway (1903–1929), late Fellow of Gonville & Caius College, Cambridge, Professor of Latin & Indo-European Philology. He lived a short distance to the south of the terminus, at 123 Lapwing Lane, Didsbury.

"At Withington enter Alexander, in outward form a patriarch, in mind and spirit a reincarnation in one person of all of Spinoza and much of Samuel Johnson."

A larger-than-life character, Samuel Alexander (1893–1924) was an ex-Oxford man and Professor of Philosophy in the Faculty of Arts. He lived at 24 Brunswick Road, Withington.

"At Derby Road a huge figure would stoop to enter. He was Rutherford. Running through Fallowfield our tramcar would pick up Elliott Smith, reputedly an Anatomist of renown, to us humanists apparently interested only in the culture of Peru."

Professor of Anatomy, Grafton Elliott-Smith (1909–1919) lived at 4 Willowbank, Fallowfield.

He accepted an invitation to become the president of the International Institute for Psychical Research which caused him some professional embarrassment.[29]

"Next Dixon, our chemist, with Horace (in Latin) sticking out of his pocket, and Lamb ("a really great mathematician" is Sir Edmund Whitaker's estimate of him)."

Harold Baily Dixon, Professor of Chemistry (1887–1922) and also, for a time, of Metallurgy (1887–1906), probably avoided sitting next to Rutherford. In 1913 Dixon lived quite close to the University in the then fashionable Victoria Park area, whereas Lamb at 6 Wilbraham Road, would have boarded considerably sooner. Horace Lamb (1885–1920) served longer as Professor of Mathematics at Manchester than anyone else.

"And finally, at the top of the rise, Tout, a stocky, thrusting figure, seeking, even on the tramcar, whom he might convince of the rightness of his latest scheme for the welfare of the University or alternatively devour."

Thomas Frederick Tout served as Professor of History (1890–1902), Professor of Mediaeval and Modern History (1902–1920) and

---

[29] For further details see Alan Gall, "From the Archives: Soul Searching and the International Institute for Psychical Research", *The Journal of the Institute of Science & Technology*, Spring 2016.

finally Professor of History and Director of Advanced Studies (1920–1925). In 1913, Tout lived at 55 Mauldeth Road, in the vicinity of Rutherford's house, so is unlikely to have been the last to board.

First World War

The group photograph (Figure 7) shows the Baumbachs on a family outing in about 1913, with glassblowing employee Fritz Hartwig (about whom, more later). In short order, their comfortable life would be shattered.

Otto's confrontation with a post-doctoral researcher called Edward Neville da Costa Andrade has been reported in various Rutherford biographies. We only have Andrade's account of what happened.

When I had occasion to call on him shortly after the war broke out, he, who knew that I understood the language, broke out into a stream of fiery German prophecy as to what the German Army would

Figure 7.   Photograph taken at Hale, Cheshire, around 1913. Left to right front: Friedrich Carl Baumbach, Gottlob Otto Baumbach with Otto Karl (later Charles) Baumbach, Lilian Baumbach (née Cowlishaw), John Charles Cowlishaw, Rudolf Nauber. Left to right rear: Nauber (died young), Otto Nauber, Fritz Hartwig. Courtesy of Philip Baumbach.

do to Britain. I answered him in the kind of way he understood, telling him to keep his mouth shut or he would find himself in trouble. The next thing was that the Vice-Chancellor of the University sent for me and told me that Baumbach had complained that I had threatened him: it was very unworthy behaviour to threaten a poor defenceless German in our midst and he must ask me not to behave in this way![30]

Otto's sons Geoffrey and Charles found this difficult to reconcile with the nature of the man they knew intimately. An article in *New Scientist*[31] that retold the supposed outburst prompted Geoffrey to write in his father's defence.

The patriotic outbursts are likely to be quite true. He was only 32 at the time and perhaps somewhat hot headed ... The outbreaks were not spontaneous however. He was not the sort of man to make rude or ill conceived provocative remarks. What has to be remembered however is that part of the strategy of conducting any war is to inflame public opinion against the enemy. During that war, damage was done to German property and shop windows were smashed, even in Manchester. My father was an enemy alien surrounded by now with a somewhat hostile population. It is not surprising that he stood his corner now and again, even if a little indiscreetly.[32]

We might conclude that, since the family's opinion is likely to be biased, Andrade's account is more reliable. However, Andrade, who adopted the name "Percy" at Manchester, did establish a reputation for being a "difficult" character.[33]

The sinking of the luxury ocean liner Lusitania by a German U-boat in May 1915 did much to engender hostility to all things German. Businesses in Manchester suffered from riotous behaviour — sometimes only on the basis that the name sounded German. As an example, most of the pork butchers in the City were German run and became the target of stone-throwing mobs.

---

[30]Edward N. de C. Andrade (1962) p. 32.

[31]Glyn Jones (1988).

[32]Letter from Geoffrey Baumbach to the editor of *New Scientist*, undated.

[33]See Frank James & Vivian Quirke (2002) for the problems caused by Andrade at the Royal Institution.

Figure 8. A box made for Otto at Oldgate camp, 1916. Courtesy of Philip Baumbach.

Rutherford tried to maintain permission for Otto to remain at the University during the war but, according to Niels Bohr, his "indiscrete" utterances led to internment.[34] The beautifully crafted wooden box shown in Figure 8 is still in the possession of the Baumbach family. Inscribed on the inside of the lid is "AUS DER KRIEGS-GEFANGENSCHAFT OLDCASTLE, IRLAND 1916" (translation: from the prisoner of war camp, Oldcastle, Ireland 1916) — a memento of Otto's "indiscretion". The camp had been a workhouse in the town of Oldcastle, County Meath, Ireland, commandeered and re-purposed in late 1914 to hold civilian "enemy aliens".

Early this morning a special steamer — the SS Duke of Clarence, belonging to the Lancashire and Yorkshire Railway Co Ltd — arrived

---

[34] Niels Bohr (1958).

from across the Channel at the North Wall.[35] *Her purpose in visiting Dublin was to convey here a party of upwards of two hundred German aliens of the civilian class, who are to be interned in concentration camps at this side of the water.*[36]

It is not known if this batch included Otto Baumbach. Many of those already there were apprehended in Ireland and the first mention of arrivals, as reported in the local press, is in December 1914.[37]

Like many ex-internees, Otto spoke little of his experiences and a popular misconception of the time was that prisoners enjoyed an all-expenses paid holiday. What befell Otto at Oldcastle can be surmised from the account of fellow inmate Aloys Fleischmann.[38] After his arrest on the night of 4 January 1916, Fleischmann spent the next thirty-nine months at Oldcastle. Stories of primitive conditions, lack of food and medical attention, the cold, a few deaths, removals to lunatic asylums and restricted interaction with other inmates does not suggest a cosy life. Above all, he stressed the debilitating effects of boredom. The camp closed in 1918 and all inmates were shipped off to the Isle of Man. In later life Fleischmann spoke to his grandchildren about the reception that the prisoners endured on arrival from a large group of women who shouted abuse and prodded them with hat-pins. The press had done well in creating great animosity with headlines such as "The Huns" Paradise — bands, concerts and picnics for Isle of Man aliens.[39]

Otto had at least two employees in addition to the Nauber brothers as apprentices. The brothers were destined for interment

---

[35] The North Wall is an area of docklands at Dublin.

[36] A cutting from *Freemans Journal* 17 March 1915 sent by Tom French, local history department at Meath County Library. Tom reports that searches for the passenger list of the Duke of Clarence have not been successful.

[37] By *The Meath Chronicle* 12 December 1914 reported in John Smith (2018) p. 19.

[38] Ruth Fleischmann (2018).

[39] Ruth Fleischmann (2018) p. 151, quoting from the magazine *John Bull*, 29 May 1915.

like their uncle, but glassblower Fritz Hartwig[40], another German, did not suffer the same fate. The other known employee, Robert Clarke, advertised his services just after war started: 'SCIENTIFIC GLASSBLOWING by competent Englishman (late with Otto Baumbach) 14 Lime Grove, Oxford Road, Manchester.'[41]

It is safe to assume that Fritz (he changed to being 'Fred') could not immediately fill Otto's shoes as several of Rutherford's co-workers have commented on experiments that could not be continued. Fritz now had the University workshop available to him and, by default, the customers. The problems that might have arisen through any remaining German connections were neatly sidestepped by calling the new enterprise The Scientific Glassblowing Company and taking on a British national as a partner. Otto never forgave Fritz for this 'betrayal', although one can readily understand the desire to take advantage of the situation.

Partnering Fritz was Alfred Frederick Edwards, assistant to professor Harold Baily Dixon. In 1904 it had been Edwards' job to find a room in the chemistry department for Chaim Weizmann.[42] Edwards may have been brought to Manchester by Dixon since both had Oxford connections. He died in 1932 and Fritz continued to run the Scientific Glassblowing Company until his own death in 1958. The company continued, under the Stuart family, and from the time of formation to the present day occupied a succession of premises: 1 Bridge Street (University premises), 12 & 14 Wright Street, 270 Oxford Road, 41 Upper Brook Street, 32 Milton Street & 163 Higginshaw Lane.

Rutherford, with the assistance of William Kay, and between war work, disintegrated nitrogen in 1917, a feat popularly known as

---

[40] A brief history of the Scientific Glassblowing Company says that Fritz Hartwig came over from Germany to join Baumbach & Co in 1908. It incorrectly (as of September 2020) gives the name of Hartwig's partner Alfred Edwards as Charles Edwards. See https://www.sciglass.co.uk/about/.

[41] *The Manchester Guardian,* 9 October 1914.

[42] Chaim Weizmann made a significant contribution to the war effort with his work on the production of acetone by fermentation, an ingredient needed for the production of the propellant cordite. He became the first president of Israel in 1949.

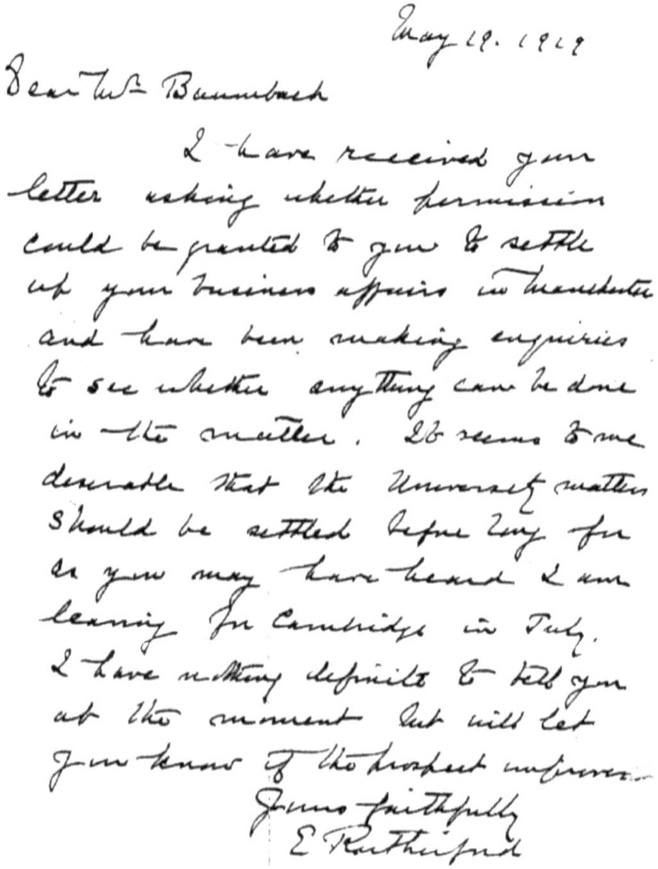

Figure 9. A letter from Rutherford embossed with his home address at 17 Wilmslow Road, Withington, Manchester. Courtesy of Philip Baumbach.

"splitting the atom". This represented his last major achievement at Manchester before being lured to Cambridge with the offer of succeeding J. J. Thomson as head of the Cavendish Laboratory. Otto had written to Rutherford, presumably while still in captivity, and received the response (Figure 9):

I have received your letter asking whether permission could be granted to you to settle up your business affairs in Manchester and have been making enquiries to see whether anything can be done in the matter. It seems to me desirable that the university matters should be settled before long for as you may have heard I am leaving for Cambridge in July. I have nothing definite to tell you at the moment but will let you know if the prospect improves.

The University Service Providers Move Out

From several accounts, it seems that the glassblower, chief technician for physics, instrument maker, and tinsmith (alias Otto Baumbach, William Alexander Kay, Charles William Cook and William Stelfox) co-operated closely and maintained some personal contact. William Kay regularly visited Otto after WWI and when Stelfox's business struggled, Otto provided the use of part of his own premises. Charles Cook moved from Manchester to Ashby de la Zouch and was visited there by Otto.

The Electro-Technics department applied to the University Senate for more space in 1919, in anticipation of rising student numbers. Attention turned to the rooms occupied by the service providers who, after all, ran private businesses and were not part of the department. Under some, presumably, mutual agreement with the University, the glassblowing facility (reformed as the Scientific Glassblowing Company after Otto's internment) moved to Wright Street. Stelfox moved into part of Otto's recently acquired premises on Bridge Street. The moves were not completed until 1926–27[43] and Chas Cook had already departed for a workshop and plush hotel at Ashby de la Zouch in about 1922.

There were varying degrees of success in these businesses. First to fall, Stelfox shareholders voted to put the firm into liquidation in 1928.[44] Chas W. Cook & Sons survived many changes of location after moving from Ashby to Birmingham, until it closed shortly after 1989. It was resurrected briefly as Chas W. Cook (MM) Ltd in 1991 under new management. The Scientific Glassblowing Company is still trading today.

---

[43] An account of the accommodation used by the service providers is given in T. E. Broadbent (1998).

[44] By an Extraordinary General Meeting on 13 December 1928. National Archives record BT31/16779/72909.

Since Otto and William Stelfox had a close connection, it is interesting to consider the tinsmiths a little further. Stelfox Ltd received its certificate of incorporation dated 15 February 1902 to take over William Stelfox's existing business as a going concern. An inventory of equipment and materials accompanied the sale agreement. So detailed is this list, in terms of accounting for every last piece of metal, file and pair of tongs in the place, trivial items of low value, that it comes as no surprise to discover that the total value of shares received by William Stelfox amounted to just £70.[45] Hardly a big sum, even in 1902. Stelfox continued to run the business but shareholders now included John Clough Thresh, Medical Officer of Heath for Essex County Council, and Bonner Harris Mumby, Medical Superintendent of the Borough Lunatic Asylum in Portsmouth. The explanation of why these medical men would be interested in a Manchester based firm of "tin bashers" is probably through Professor Sheridan Delepine at the University. Delepine and Dr Thresh co-authored articles on public health and Thresh had devised an incubator, a piece of equipment that could easily be made by Stelfox.

The Inter-War Years

Otto returned to a Manchester devoid of German pork butchers. The success of Otto Baumbach & Co had paid for a comfortable lifestyle, now much reduced. Weizmann wrote from London to settle an outstanding bill and enquired "I shall be glad to hear from you whether you are continuing your former work" (Figure 10). Indeed he was, but restrictions placed on enemy aliens made running a business far from easy. They were treated almost as bankrupts, with control in the hands of the Custodian of Enemy Property until 1925. The house in Hale had to be sold and residence taken up in the less salubrious district of Old Trafford. Otto bought 403 Chester Road in 1919/20 where the business now operated as J. C. Cowlishaw, Thermometer

---

[45]The file on Stelfox Ltd is held at the National Archives, reference BT31/16779/72909.

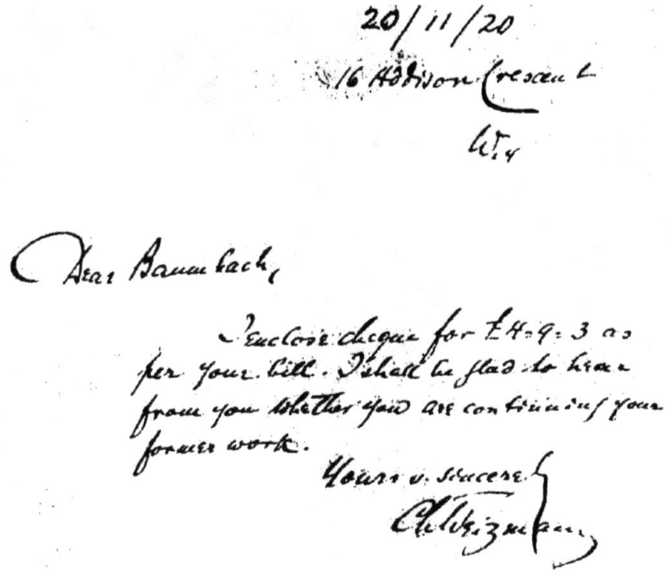

Figure 10.   A belated payment from Chaim Weizmann. Courtesy of Philip Baumbach.

makers, and where the family also lived. After a short time alternative living arrangements were taken up at "Rivington", Irlam Road, Flixton, Geoffrey's place of birth in 1921. J. C. Cowlishaw Ltd received its certificate of incorporation on 15 July 1925. The goodwill of the business was somehow precisely valued at 199 pounds, eight shillings and eleven pence!

Thermometers were the bread and butter of Cowlishaw's activities but a limited market for these products posed a problem. Then came a lifeline in about 1924, an opportunity to supply the Manchester based firm of Mather & Platt Ltd with the glass bulbs used in automatic fire protection sprinklers. Teething problems were considerable. Geoffrey described the process of forming a glass blank on a "drawnout machine", converting it to a semi-molten blob of about the correct shape and size, and inserting it into a mould (having previously connected a syringe to inject air).

It tends to sound reasonably simple when so described, but was in fact an extremely skilled operation. A new operator being trained

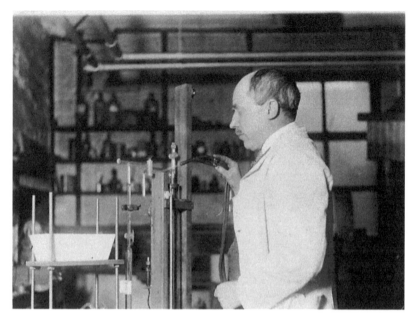

Figure 11.   Otto at 42 Bridge Street 1928. Courtesy of Philip Baumbach.

could practise for several hours each day and still only be able to produce the odd good moulding after 6 months or so. On the other hand, a skilled operator would produce upwards of 150 good mouldings each working day.[46]

Not long after the start of bulb production came a move to 42 Bridge Street (later Bridgeford Street), back to being close to the University. Figure 11 shows Otto at work two years after moving in. Production for Mather & Platt continued alongside the construction of special apparatus for university and other customers.

Rutherford had not forgotten Otto's skills and wrote from Cambridge in 1922: "Dear Baumbach, I shall be glad if you will make within the next fortnight & send to me a pump of the enclosed design — similar to those you made for me in m/c [Manchester]" (Figure 12).

---

[46] Geoffrey Baumbach (1994) p. 3.

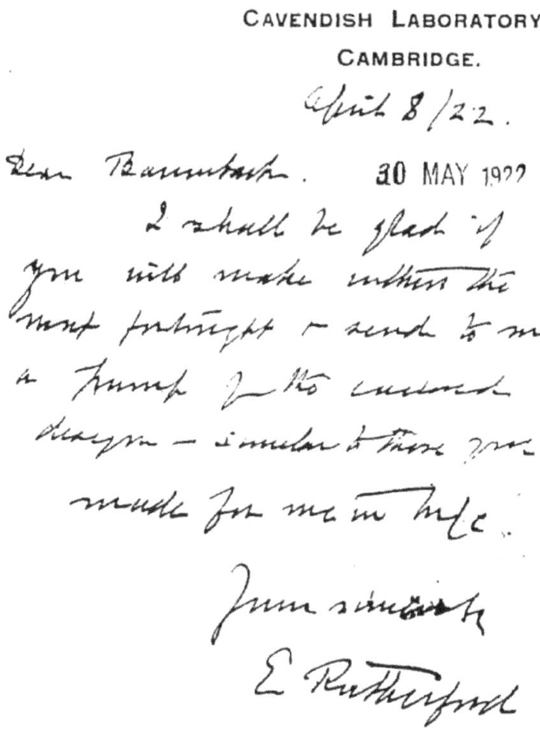

CAVENDISH LABORATORY,

CAMBRIDGE.

April 8/22.

Dear Baumbach. 30 MAY 1922

I shall be glad if you will make within the next fortnight & send to me a pump of the enclosed design — similar to those you made for me in M/c.

Yours sincerely

E. Rutherford

Figure 12. Rutherford still had need of Otto's services until the appointment of a glassblower by the University of Cambridge.

Other universities provided Otto with custom. Edmund Bowen described the primitive state of the Balliol-Trinity Laboratories at Oxford in the 1930s:

The main laboratory was heated by an enormous Bunsen burner placed below a wide, sinuous cast-iron pipe on one wall. Thermostats were fitted with toluene-mercury regulators controlling the supply of gas to a heating jet. There was no workshop, glass-blower, or technical assistance; special glass apparatus was sometimes ordered from Otto Baumbach of Manchester ..."[47]

---

[47]E. J. Bowen (1970) p. 235. Edmund (Ted) Bowen focused his research on chemical fluorescence, for which he was elected to Fellowship of the Royal Society in 1935.

Bernard Lovell, later Professor Sir Bernard Lovell the director of Jodrell Bank observatory, arrived at the University of Manchester to join W. L. Bragg's staff in 1936 and has written of the experience:

I tried, but failed absolutely, to interest myself in Bragg's research and made the bad mistake of trying to restart my thin film work in a place with no facilities and with only a poor quality commercial glassblower several streets away.[48]

The location of a glassblower several streets away may fit in with Otto's business operating from 42 Bridge Street, but the comment about the standard of workmanship is totally at odds with the evaluation of many others, even the abrasive Andrade.[49] This required some clarification and Sir Bernard responded to my enquiry:

As regards Otto Baumbach, he is undoubtedly the glassblower referred to in my article. I fear that my unenthusiastic reference to him is most unfair. I had just come from Bristol where Burrows[50] had succeeded in making the most complex Pyrex high vacuum equipment for my research on the deposition of thin films of the alkali metals and this type of apparatus was beyond Baumbach's experience. I visited him a few times when he worked in a most depressing area a few streets away from the Manchester Physics Department.[51]

The Second World War and After

The inhabitants of boarding houses on the Mooragh Promenade, Ramsey, Isle of Man, received a rude shock on Monday 13 May 1940 when a policeman served them with an official notice to vacate. Two weeks later the houses formed the basis of the Island's first Second World War internment camp and Otto's home for at least part of his second incarceration. It is believed that he spent time in one or more other camps. Sometime during internment he paid to have

---

[48] Bernard Lovell (1987) pp. 156–157.
[49] Edward N. de C. Andrade (1964) p. 101.
[50] J. H. (Johnny) Burrow, glassblower at the University of Bristol. See, for example, the biography of Bernard Lovell by Saward (1984) pp. 24–25. He was a founder member of the British Society of Scientific Glassblowers.
[51] Letter, Sir Bernard Lovell to the author 27 March 2003.

Figure 13. A portrait of Otto painted during WWII internment. Courtesy of Philip Baumbach.

his portrait done, an impressive result using a scrap piece of wood (Figure 13).

World War Two had begun on 1 September 1939 and on the 18th Geoffrey Baumbach joined his brother as a director of J. C. Cowlishaw Ltd. Thus, with the absence of Otto, the responsibility of running the business fell on the shoulders of Geoffrey, not yet nineteen years of age, and his elder brother by eleven years, Charles (born Otto Karl). At that time there were over twenty employees to supervise.

Mooragh Camp held a number of Finnish prisoners without regard to their affiliations. Several violent confrontations resulted between pro- and ant-Nazi Finns, culminating in the murder of Nestor Huppunen in 1943. By the time of Otto's release on 19 September 1944, he had spent a total of nine years in captivity over the two wars. According to Geoffrey, the experience did not leave him embittered and he soon resumed his old work. One of his first acts on returning home was to construct a complex glass manifold, an impressive performance after five years without any practice. Otto finally obtained a Certificate of Naturalisation dated 28 October 1946.

In the same year as Otto's naturalisation, Charles interviewed school levers for jobs as trainee glassblowers. A selection made, fourteen-year-old Stanley Taylor joined Cowlishaws as an apprentice the following Monday after being shown around the works.

I met Otto for the first time that day, and although he spoke English there were certain words he could not pronounce, and one of the things he asked me was if I had a farter. I was dumbfounded. I didn't know what to say but Charles came to my rescue and explained that he could not pronounce 'TH' and he was asking about my father.[52]

Stan Taylor kept a Cowlishaw catalogue, undated but probably pre-1960. Some of the products are shown in Figures 14–16.

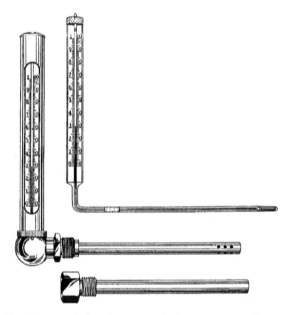

Figure 14.   Cowlishaw tubular, brass cased thermometers. Courtesy of Stanley Taylor.

---

[52]Letter from Stanley Taylor to the author 12 March 2008.

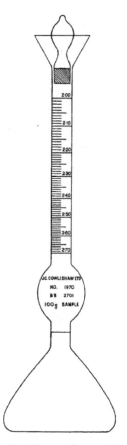

Figure 15. Cowlishaw Rees Hugill specific gravity flask. Courtesy of Stanley Taylor.

Thermometer housings were made by Mallan Engineers Ltd, originally an independent company but they owed money to Cowlishaws and were taken over to settle the debt. Stan worked his way up from apprentice to become a director of Cowlishaws, appointed 16 July 1976. Another employee to become a director, Brian Arthur Hosie, came straight from the Army in 1954 to set up an engineering workshop.

At the start of the 1960s, the premises at 42 Bridgeford Street (previously named Bridge Street) were compulsory purchased by Manchester Corporation for, as Geoffrey put it, a "song". A location

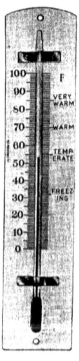

Figure 16.   A humble household thermometer. The scale is of glazed earthenware.
Courtesy of Stanley Taylor.

on Pilling Street (later renamed Peary Street) proved to be suitable
and building commenced.

> *We received very shabby treatment at that time from Manchester
> Corporation. Even threats that they would throw our possessions into
> the street because the progress of our almost completed new building
> was a little behind schedule and we were thus impeding progress of the
> new school which was to occupy the site which we were vacating.*[53]

About sixty people were eventually employed hand-making bulbs
until it became uneconomic to continue production. The firm moved
into electronic circuit assembly work as the requirement for sprinkler
bulbs, thermometers and glass apparatus started to fall. In 1979,

---

[53] Geoffrey Baumbach (1994) p. 5.

after the difficult task of serving redundancy notices on employees of long standing, it became obvious that the only sensible answer was a voluntary liquidation. An official winding-up followed in 1982.

Otto Baumbach lived to see the opening of the new factory, partly financed from the sale of his house "Four Gables" at Mobberley. By this time, mental deterioration had set in. He went to live with Geoffrey at "Pinewood", Mobberley. The family moved to Middleton but Otto only lived for a further ten days, dying there at age 83 on 15 May 1966.

A letter to Geoffrey Baumbach dated November 1997 begins: "This letter is one which I have been thinking of writing for the past twenty years or more, in any case, one I should have written at least fifteen years ago."[54] So began seventeen pages of reminiscences about working at Cowlishaws, a joint effort by Raymond and Clifford Ford. The contents were a heartfelt appreciation of the benign conditions of employment and the esteem in which the Baumbachs were held. Raymond had a computer and printer at his disposal but wrote: ". . . I feel longhand will convey the sincerity and depth of feelings in the words I am writing."

The letter spoke of paid sick leave and generous holiday entitlement, uncommon for shop floor workers of the time. Otto, Charles and Geoffrey ran the business in a paternalist way but occasional reprimands were administered and accepted. Shortly after joining, Clifford had occasion to assist Otto with the construction of a glass tube. This required Clifford to blow at one end with just enough pressure for the desired outcome.

Cliff, crouched on the floor, got the request to blow, which he did but not knowing how hard and, apparently, not hard enough. A further shout from your Dad[Otto], "blow", Cliff blew again. There was a further agitated shout of "blow". This time Clifford overdid it and whatever your Dad was making was spoilt and your Dad whacked Clifford on the head. Great stuff!

---

[54]Original letter in the keeping of Philip Baumbach. Mention of "fifteen years ago" is a reference to the closure of Cowlishaws.

Stan Taylor confirms the generally easy-going regime: "I enjoyed all my working life with them, I even met my wife there, it was like being part of a large family. They were not just our employers. All the people who worked there felt the same".

Conclusion

During the period 1907–1914 the glassblowing skill of Otto Baumbach was a timely resource that enabled Rutherford and his co-workers to employ considerably more complicated apparatus in their experiments than would otherwise have been the case. Otto continued after 1919 but with the emphasis on work of a commercial nature. However, his reputation in the academic world ensured further patronage from university departments for some time after.

Otto's second spell of internment came as a bitter blow for a man who had already paid for his earlier indiscretions and had given no further offence to his adopted country. Yet despite the deep hurt that this caused, he remained an Anglophile. Friends, family and employees alike remember him as an intelligent, hard-working man who was modest enough to look back on his life without thinking he had done anything special.

Acknowledgements

Thanks go to following who have helped with this project over a number of years, and some more recently.

The late John Geoffrey Baumbach, son of Otto, and Philip Baumbach, grandson

Yvonne Cresswell, Manx National Heritage

Tom French, local historian at Meath County Library

Brian Hosie, ex-J. C. Cowlishaw Ltd

The late Keith Papworth at the Cavendish Laboratory, University of Cambridge

Stanley Taylor, ex-J. C. Cowlishaw Ltd.

Bibliography

Andrade, E. N. da C., *Rutherford and the Nature of the Atom* (New York: Anchor Books, 1964).

Andrade, E. N. da C., "Rutherford at Manchester 1913–14", in J. B. Birks (ed.) *Rutherford at Manchester* (London: Heywood & Co, 1962), pp. 27–42.

Bohr, Niels. "The Rutherford Memorial Lecture 1958: Reminiscences of the Founder of Nuclear Science and of Some Developments Based on His Work", *Proceedings of the Physical Society*, 78/6 (1961), pp. 1083–1115.

Baumbach, Geoffrey. *Brief Review of G. O. Baumbach* (unpublished manuscript, 1994).

Bowen, Edmund J., "The Balliol-Trinity Laboratories, Oxford 1853–1940", *Notes and Records of the Royal Society of London*, 25/2 (1970), pp. 227–236.

Broadbent, T. E., "Electrical Engineering at Manchester University: 125 Years of Achievement" (The Manchester School of Engineering, University of Manchester, 1998).

Burt, F. P. and Edgar E. C., "The Combining Volumes of Hydrogen and Oxygen", *Philosophical Transactions of the Royal Society of London. Series A*. Vol. 216 (1916), pp. 393–427.

Campbell, John, *Rutherford: Scientist Supreme* (Christchurch, New Zealand: AAS Publications, 1999).

Dahl, Per F., *Flash of the Cathode Rays: A History of J J Thomson's Electron* (Bristol & Philadelphia: Institute of Physics Publishing, 1997).

Day, David T., *Mineral Resources of the United States: Calendar Year 1905* (Washington: Government Printing Office, 1906).

Dixon, Harold B. and Edgar, E. C., "The Atomic Weight of Chlorine: An Attempt to Determine the Equivalent of Chlorine by Direct Burning with Hydrogen", *Philosophical Transactions of the Royal Society of London. Series A*. Vol 205 (1906), pp. 169–200.

Fleischmann, Ruth, "Aloys Fleischmann: Bavarian Musician & Civilian Prisoner of War 1916–1919" in Tom French (ed.) *Oldcastle Camp 1914–1918: An illustrated History* (Meath County Council, 2018), pp. 125–166.

Frederick-Frost, Kristen M.,"An Artifact Based Study of Henry Moseley's X-Ray Spectroscopy", MSc Dissertation (University of Oxford, 2012).

French, Tom (ed.) *Oldcastle Camp 1914–1918: An illustrated History* (Meath County Council, 2018).

Heilbron, J. L., H. G. J. Moseley: *The Life and Letters of an English Physicist 1887–1915* (Berkeley, Los Angles & London: University of California Press, 1974).

Holmberg, Peter, "Lars William Öholm — Letters from Manchester 1913" in Malcolm Cooper (ed.) *Newsletter of the History of Physics Group of the Institute of Physics (UK & Ireland)*, No 34, December 2016.

James, Frank A. J. L. and Quirke, Vivian, "L'affaire Andrade or How Not to Modernise a Traditional Institution" in Frank A. J. L. James (ed.) *The Common Purposes of Life: Science and Society at the Royal Institution of Great Britain* (Aldershot: Ashgate Publishing Company, 2002), pp. 273–304.

Jones, Glyn, "The Phantom of the Atom", *New Scientist*, 28 January (1988), pp. 56–59.

Lovell, Sir Bernard, "Bristol and Manchester: The Years 1931–9", in Rajkumari Williamson (ed.) *The Making of Physicists* (Bristol: Adam Hilger, 1987), pp. 148–160.

Mendelssohn, K., *The Quest for Absolute Zero* (London: Weidenfeld & Nicolson, 1966).

Rowland, John, Ernest Rutherford: Atom Pioneer (London: T. Werner Laurie, 1955).

Saward, Dudley, *Bernard Lovell: A Biography* (London: Robert Hale, 1984).

Smith, John, "Oldcastle Camp 1914–1918: An Overview" in Tom French (ed.) *Oldcastle Camp 1914–1918: An illustrated History* (Meath County Council, 2018), pp. 7–52.

Rutherford, E. and Royds, T., "The Nature of the α Particle from Radioactive Substances" *Philosophical Magazine*, 6/17 (1909), pp. 281–286.

Wende, Peter, A History of Germany (New York: Palgrave Macmillan, 2005).

Westaway, Jonathan, "The German Community in Manchester, Middle-Class Culture and the Development of Mountaineering in Britain, c. 1850–1914", *English Historical Review*, 124/508 (2009), pp. 571–604.

Wilson, David, *Rutherford: Simple Genius* (Cambridge, Mass: MIT Press, 1983).

Winterburn, Emily Jane, 'Rutherford at Manchester 1907–1919', MSc Dissertation (University of Manchester, 1998).

Woolleys of Manchester: A Record of 150 Years in Pharmacy (Manchester: James Woolley, Sons & Co Ltd, c.1946).

Other references are contained in the footnotes.

This chapter is a revised and much expanded version of articles which appeared in *Science Technology* December 2004, the *British Society of Scientific Glassblowers' Journal* April 2004 and in the *Newsletter of the History of Physics Group of the Institute of Physics* January 2008.

Chapter 11

Stephen Moehr: A Working Life

Stephen W. Moehr

Former Chief head of Scientific Glassblowing York University 1966–2007
Former Chief Examiner British Society of Scientific Glassblowers
Training Consultant 2008–2011

I was at Allerton Grange school aged 14 and I had no idea what I wanted to do for a career.

The deputy head asked me to make a model demonstrating the power of water. He said it looked particularly good and thought I was good with my hands. He told me there was a job going at Leeds University in the Chemistry Department for a Laboratory Technician and I got an interview. On learning I was good with my hands they said there was a job going in the glassblowing department which they thought might suit me better. I was taken down to the glassblowing workshop and introduced to the boss, Harry Butler. The post had occurred because the glassblower, Barry Davis, and been called up. I was told I might not be suitable as only a few were able to do it. I was told I could go in and give it a try to see if I liked it and to see if I had the potential. I would not get paid. Well, my father had me going in every day for two weeks, remember I was only 14. Harry Butler, my boss, had me pulling points in soda glass until they spun true and were thick enough to hold their own weight. I was offered the job starting on my 15th birthday. I must have done nothing else but pull points for a couple of weeks until they were perfect. I then had to practice test tube ends and bends, still in soda glass.

It was very boring. However, there was another apprentice called Martin Watson. We got on well and had lots of fun teasing each other. Later, David Flack joined us, a nice guy, but unfortunately, he died tragically in a motor bike accident.

Eventually I was allowed to work with borosilicate glass, which I found strange at first. I was not allowed on the lathe until I could make most things by hand. Harry Butler would get me to blow for him when he was working on the lathe. I remember once he was doing quite a difficult job and the flames were very noisy. He shouted at me to blow, then "harder lad", then he shouted "stop", which I did not hear, and I blew like mad and blew a hole in his job. I think he did his own blowing after that.

They were long days. I had to leave my house in Moortown Leeds at 7.30 am and get one bus into town then another bus to the University to start at 8.45 am. We did not finish until 5.30 pm and we also had to work Saturday mornings.

As an apprentice, I had to make the coffee in the morning and tea in the afternoon. This was a welcome break from glassblowing. I took completed jobs back to the labs and went to the general stores for bottles of oxygen. I fell for all the tricks, like going for a long stand, they kept me waiting ages for that, a box of sky hooks, etc.

I was now on an official five-year apprenticeship. I was practicing straight joints, bends, internal seals, cold traps, condensers and eventually three-stage diffusion pumps. It was about this time the BSSG had just started up and they had their first day's Symposium at the Leeds Metropol Hotel. We started the North Eastern section, which I attended, which included five glassblowers from ICI Harrogate headed by Rex Eustance.

Harry Butler was an extremely hard taskmaster. If he was in a mood, he would tell me that what I had just made wasn't good enough and throw it in the bin instead of showing me where I had gone wrong and how to improve it. Fortunately for me, Barry Davis came out of the army. He was an excellent, very skilled glassblower and he took over my training.

Barry eventually left for the position of Chief Glassblower at York University. There was just one Science building, Chemistry.

It housed both Biology and Physics. I was ready to move on as I was always the "lad" under Harry Butler, I was offered a job at Hull University, but that would have meant me staying away from home, which I didn't fancy. Then Barry Davis informed me that York University was looking for a Physics Glassblower and I should go for that as I would be designing my own workshop when the building was completed. I took the job, but it was quite different glassblowing to that I was trained for. Fortunately, for the first two years I worked along with Barry in the Chemistry Department until the Physics workshop was built, much of the work was making valves working with C9 (W1) glass and some very large Flowing-after glow tubes. I also had to make Argon Ion silica lasers working with Professor Heavens, a pioneer of lasers in those days. There were no computer-aided drawings then, I do remember a Professor from Australia tearing a bit of cardboard off a box and drawing a laser he wanted me to make.

I still lived in Leeds and traveled to York each day on my Honda motor bike, a journey that took 40 minutes. Pamela and I got engaged and put a deposit on a property in a new housing development in Woodthorpe Garden Village, York. We married in 1967 and moved into the new house, a lovely time. I changed my motor bike for a reliant three-wheeler van which I had for three years before I passed my car test. I remember Barry who was now at his own business, York Glassblowing Services, asking me if I could take him a big Dewar of liquid nitrogen. I shared driving to and from work with the head of Biology. He refused to sit in the front and hold the Dewar, so I wedged it as best I could, and he sat in the back. On the way to Barry's, the car in front suddenly stopped and the Dewar fell over. Well there was vapor everywhere, pouring out of my fibre glass car and my rubber mats in the front were banging loudly. Wally was shouting "let me out" as cars behind were reversing away. We were lucky no one called the fire brigade. When we got to Barry's he complained I had not brought him enough. I showed him my shattered rubber car mats.

I moved into my new workshop in the Physics Department. I had a specially adapted 8 inch bore Heathway lathe, which as well as doing centrifuging also had a separate DC Tram motor going terribly slow for coil winding. It was a monster. Much of the equipment was bought through Jencons Scientific and their representative, John Beeson. I was happy there and I took on my first apprentice, Richard Hall. We worked well together, and he eventually left for a job in Swindon. Two of the Professors for whom I did a lot of work left, which took about 50% of the work from me. I was then asked if I would like to work in analyzing the York Minster glass. The University had a contract from British Nuclear fuels for us to find a glass, an incredibly old glass that had not weathered. They were looking at the possibility of encapsulating waste fuel in glass and burying it in the seabed. It was an interesting job as I had to go to extremely high places in the Minster. We were provided with samples of different old glasses on which I polished a small corner of the glass, using only diamond polishing materials. The glass was then analyzed by non-destructive X Ray fluorescence.

About 1970 I was asked to go on the Board of Examiners for the North Eastern section. I had to prove to the board that I had the experience to be a member. Members of the board worked so hard in formulating those examinations. My first job was as Competition Secretary, a job I really enjoyed. The Master's examination was being written and my module was Optical Polishing and Glass to Metal Seals. At each meeting we had to produce questions and answers for each of the examinations. Norman Collins, the secretary worked extremely hard. A special cup in his name is awarded to the member whose entry of scientific glass apparatus is considered by the Board of Examiners to be the most outstanding example of craftsmanship.

I was Competition Secretary and eventually Chairman. As I remember, Tom Maple was the first Chairman I worked with along with Norman Ellis, Rex Eustance, Bert Chappell, Keith Holden, John Turnock, Fred Porter, Fred Morse and Mike Lock. Many hours were spent by the board setting out the examinations, which are still used today.

Bert and I had the privilege of examining Graham Reed for the first and so far only Master's examination by qualification.

Eventually, the glassblowing dried up in the Physics Department, but there was a growing need for more optical polishing. I was sent to a firm in Brentwood, Electro Optic Developments, where I was taught various polishing techniques. I enjoyed that, polishing to extremely high tolerances.

I was asked to form a central glassblowing service in the Chemistry Department doing glassblowing for all three science Departments.

I took on Brian Smith who was another apprentice taken from Leeds University who worked in the Biology Department.

Barry had left the Chemistry Department to start a new business, Northern Scientific, doing both glassblowing and acting as an agent for Edwards High Vacuum Pumps. Eventually, they split the company and Barry started York Glassblowing Services, which became Yorlab Ltd. It was run by Barry's son Alex. Jeff Clarke continued Northern Scientific (York).

I reluctantly took early retirement in 2007 as I loved my job. I was missing male company and the comradeship. Two years later I was asked to go back to work two mornings a week to train the new apprentice Abby Mortimer. That was just what I wanted. I really enjoyed that; Abby took all the BOE examinations leading to the Standard of Competence. She was examined by William Fludgate and Brian Moore. Abby passed the examination with the highest marks recorded. I felt that was a fitting end to my working life.

Chapter 12

The Wherewithal

Terri Adams

Senior Scientific Glassblower, the University of Oxford, Oxford, UK

Mastering the manipulation skills and dexterity required to support an effective research scientific glassblowing service is certainly a challenge and something not everybody is cut out to do. One could say that many of the skills needed in the manufacture of such sophisticated glassware must become instinctive — from the sound of the lathe at the required speed to the feel of the glass ahead of a bend and the colour of the flame. It is no mean feat to achieve a level of competence in manipulation and accuracy. Mastering these fundamental skills would set anyone in good stead to pursue a successful and rewarding career in scientific glassblowing.

I cannot remember the last time I was asked to make a mercury diffusion pump; there is no need to these days as fantastically quick, clean turbo boosted pumping stations are available and mercury-free — just one of the somewhat inevitable evolutionary alternatives to glass as a research material, a fact responsible for the demise of many, dare I say, more fragile innovations in scientific research practice. Critically, the skills and knowledge needed to successfully produce a working diffusion pump are invaluable in a modern workshop. As are skills such as the ability to make stopcocks, joints and flanges accurately and effectively from scratch. Also, the somewhat unconventional practices such as suck seals, diaphragm seals and blowing into rod are all skills worthy of a place in a modern workshop's skills arsenal, so justifiably integral to an advanced training program. The internet is a great modern resource for practical training workshops, but there is no substitution for standing alongside a master at work and seeing up close the subtleties of the craft and having face to face critique/instruction.

However, for me the more difficult part of the job comes when heading up a research glassblowing service. In my mind, the key component in success is extracting sufficient information from the customer to give you the best shot at success in designing a piece of apparatus which will be fit for purpose. Followed closely by managing people's expectations (and egos) and delivering an efficient, cost-effective service. If you are fortunate, your customer will arrive with a sketch/drawing, or a research paper for reference, but all too often that will not be the case.

If you are successful in achieving a full brief, you can pen your design and seek approval from the customer. Next you need to be able to reason your way through and plan the manufacture process to successfully produce a working vessel. If you get the assembly/construction process incorrect, things can go catastrophically wrong very quickly. Things can suck in or be blown out. Side arms can be knocked off or prevent the job from being removed from a lathe, that kind of thing. This is something which can't really be taught, it can be expanded upon and fine-tuned with experience, but you either have the aptitude for it or you don't.

There are numerous established calculations and equations, information tables and references that can support us in the design process, but a comprehensive understanding of the versatility, limitations and compatibility of the materials we are using is key. You really do not need to be a scientist, your customer is the expert in that field, but you do need to be able to extract from them the extreme parameters of the proposed piece, get an overview of the science taking place within the apparatus and any relationship it must have to associated pieces of equipment or any forward journey it may be subjected to.

Do not be afraid to push the boundaries and question your customer. It would not be uncommon for researchers to have put in the minimum amount of effort in securing their design or experimental plan before they approach the glassblower. A few standard questions would soon confirm this as the case before they are dispatched to do more preparation ahead of their return. You will find they can produce some magnificent drawings with a computer (or pencil and ruler if they know how to work them!), but they all too often have no perception of what is possible or impossible in glass. This is where you are the expert, and your customer will respect this.

I found first setting foot in Oxford University's Chemistry Department very intimidating, after all these were some of the best brains in the world, but I very soon realized that "their" perception of me had much the same effect — apparently the prospect of a consultation with me could intimidated them, too... and not just because back then I was one of very few women in the department. I am straight talking, so if something doesn't make sense or sit well, I will say so or question it, in the politest manner, after all it's not what you say, it is how you say it, right? And a more helpful sole you would be hard pressed to find... in a glass workshop! I try never to turn a job away, but I do appreciate my limitations so would resolve to find an external solution if not an alternative solution to the issue. In short, I am helpful, friendly and respectful while remembering a little bit of humility goes a long way. I enjoy my job and an effective interaction with co-workers at all levels within the organization is a major part of my happiness and successful integration.

There was at first a completely unfounded sense of having to prove myself. This was Oxford after all and I had learned from experience that praise is seldom offered, assume you have done a good job, but rest assured you will certainly be pulled up about it if you have not.

Around 12 months in, I had an overambitious Post Doc who tried to demand I drop everything to prioritize his job as he worked for the Head of Department. I politely explained that it wasn't the way the workshop operated but I would try to fit his job in as soon as practicable — it was/is a very busy workshop. Needless to say, he left somewhat indignantly with a bee in his bonnet only to return a short time later with the HOD in tow. Now this tactic was new to me and at that time I was still fairly new to the department, so I'd be lying if I said I wasn't a little uneasy at the prospect of a confrontation with the HOD.

So, the Post Doc had run to the HOD and regaled his tale of the impertinent glassblower. The HOD, hands anchored firmly behind his back, strode over to my bench and began... "So, I hear you are not willing to prioritize this job for my Post Doc, is that correct?" I drew breath before confidently saying "that is correct, there was no justification for prioritizing his job". I then held my breath and waited.... . But to my utter disbelief the HOD turned to the Post Doc and calmly said, "Post Docs are ten a penny, good glassblowers are terribly hard to find now f**k off back to the rock you crawled out from under and let us both get on with our jobs!", he then apologized for the interruption, politely thanked me for my time and left. Now I had the confirmation and approval I needed to thrive and have been supported, rewarded and achieved regular investment in the workshop and career development throughout my time here.

In this modern competitive world, time and space are money, so it can be challenging to keep a balance on the economy of time with the associated cost, investment in skills and the longevity and relevance of the service. Remember, when it comes to workshop management, 'mending and making do' as an economy along with keeping your head down and just getting on can be easily misconstrued as a lack of interest and investment in the service. So, tell people who you are, what you can do for them and encourage them to invest time and money in you and the workshop.

Plasma Sculpture: The Art of Scientific Glassblowing

While training, I was actively encouraged to be creative and artistic, the logic therein being, if I could make a piece of tubing or rod look like a recognizable object, then I must have a reasonable understanding of how to successfully make the glass move.

Well, I have continued to invest time, energy and money in artistic pursuits utilizing my scientific skills and knowledge in the hope of achieving something unique and pleasing to the eye. I discovered a flare for, and a love of producing, hollow blown, often life-like or working sculptures. This artistic ability was not lost on my employers and resulted in this complimentary skill being successfully employed in scientific research in the form of scale models often working from MRI scans to produce person-specific models for the likes of stent research and teaching models.

Personally, I have never been a fan of working with coloured glass — it requires far too much patience to maintain the colour quality, so I looked to science to produce an alternative colour effect to employ with my sculptures which would be both pleasing to the eye, intriguing to the mind and set them aside from traditional glass sculptures; I decided to settle on the use of plasma.

A passion of mine has always been to promote the skills I have learned as a scientific glassblower. It has been sad to have repeatedly received unlimited praise and accolades for the various token, quickly made glass trinkets or donations I've been asked to gift for charity events when the sophisticated pieces of scientific glassware that I've labored over for hours go relatively unpraised by the end user... but the scientific stuff is my job after all — not everyone can make a pig with a 5p inside it, which I'm told is amazing!! In my experience members of the public seldom get to see or appreciate scientific glassblowing, but those that have are mind blown. So I've made it my particular mission to combine my scientific glassblowing skills and associated techniques to create elaborate, detailed art utilizing sophisticated scientific techniques to further enhance them and so as to promote scientific glassblowing, to educate, to spark interest and to generally amuse.... but hopefully amaze.

My weapon of choice in so doing.... The art of plasma sculpture.

Single electrode plasma sculpture

The first plasma lamp having been invented in 1894 by Nikola Tesla, it has been known for over a hundred years that under the right circumstances noble gases give off light when excited by an electric current. They can be energized for long periods of time without breaking down or combining with other containment elements. This makes them ideal for making plasma sculptures.

Relatively rare in the atmosphere, noble gases are gaseous elements occupying group 18 of the periodic table — also known as inert gases. They are the most stable elements due to having the maximum number of valence electrons their outer shell can hold. Therefore, they rarely react with other elements since they are already stable. The colors emitted are dependent on the gases used. Commonly used are mixtures of neon, argon, xenon and krypton. These gas mixtures can be doped with the addition of some traces of molecular gases such as mercury vapor, iodine and bromine, nitrogen and oxygen, which may be dissociated by the plasma to create additional filament effects.

In principle, anything you can create hollow could be made to glow beautifully in the dark. A plasma sculpture is a clear glass container filled with a mixture of various noble gases at nearly atmospheric pressure energized by a high voltage current to create an interactive, fluid lighting effect.

The technique I use is a process remarkably like that for making traditional neon signs, but instead of the normal two electrodes and potentially messy looking wires, you can use a single electrode (or in smaller pieces no electrode) which can support much more sophisticated, complex designs. It requires a high frequency transformer or neon 'driver' which, when connected to the electrode of the sculpture and energized, passes high voltage through the electrode that excite the atoms to glow. This high-rate electrical oscillation causes electrons from the gases to fall off. This leaves the positive ions, which gives the gases beautiful colors. Due to the partial vacuum inside the sculpture, the electric tentacles can be easily seen. Normally, electric current is invisible; however, ions of the noble gases react to the outflowing electrons causing them to glow in various colors (depending on the gas type) by emitting large numbers of photons.

Adding to the beauty is the interactive nature of a plasma sculpture. Touching the sculpture draws a colorful strand of light to your finger. It is like creating your very own bolt of lightning from the electrode to your finger. This phenomenon occurs because of the conductive properties of the human body. When the glass is touched, you create a discharge path with less resistance than the surrounding glass and gas mixture with often incredibly beautiful results.

In terms of the actual glass sculpture design, your imagination and skills are your limit. It could be that you have chosen a single volume such as a blown hollow figure, a cylinder or bulb using one gas mixture throughout. Or it might be a more ambitious piece which incorporates integral sealed pockets of different gas mixtures or even other elements such as phosphor painted figures or uranium glass features. Be mindful that the sculpture might be touched. It will need to be sealed off and the 'tip off' point will be visible, also the electrode will need to be connected to the driver and subsequently to the mains. Personally, I try to get creative with my design so as to hide these otherwise potentially ugly elements within the base.

The vacuum system

For filling my sculptures, I have built myself a filling manifold (Figure 1).

Figure 1.   Filling Manifold.

- It is made using medium walled borosilicate tubing.
- The position of the HV stopcocks allows me to introduce measured amounts of gas into the manifold and operate the filling and gas handling sides of the manifold independently of each other and without losing pressure in the whole system.
- The gas cylinders (lecture bottles) are connected to the manifold by a compression fitting (I use SVL TORION) for ease of changing/removal for storage.
- There are two stopcocks for each gas port with a 2 cc volume between them. This allows me to regulate how much gas is released into the manifold; a controlled entry for filling while minimizing the potential waste from overfilling.
- I have a manometer so I can monitor the gas pressure reasonably accurately during filling and to keep track of successful gas mixture or 'recipes' so I can easily and confidently reproduce the colors and effects should I feel so inclined.
- There is a large bore HV stopcock for ease of pumping and isolation between the manifold and the connection to the vacuum pump.
- Critically, there is an air vent on the main manifold which allows for safe venting of the system to protect the pump.

Filling

The first step in filling is to connect your sculpture to the manifold. Be sure to have a constriction in the tubing close to your sculpture to facilitate sealing off. Remember that this may be in plain sight on the finished piece so allow for that in your design. It is critical for success and longevity that the sculpture is leak free so leak test before you fill.

Once you are certain you can proceed successfully, the next operation is to drive off any moisture and impurities from the electrode and to remove the air, dust and other impurities that can cause the illumination to fail. To enable me to do this I have opted to connect the sculpture inside an annealing oven. This is in part because most of my sculptures are of a not insignificant size! Heat to a temperature of 150°C while the vacuum pulls — do this for around 15 minutes. If you do not have access to an oven or if you have a smaller piece, you can effectively warm the sculpture with a hairdryer.

Once sufficient vacuum is achieved the stopcock between the manifold and the pump is closed. Next controlled amounts of noble gas can be introduced via the two stopcocks, 2cc reservoir system. Some people prefer to flush the sculpture with gas then evacuate again before the final fill — it is a personal choice and noble gas volumes are expensive to purchase. The color and intensity of the final illumination are controlled in part by gas pressure within the sculpture, hence the manometer. To see the progress or evolution of your fill glow you can energize the gas via the electrode during filling. Too much pressure however will result in the sculptures glow to cease, then you will need to remove some gas via the vacuum pump.

When you are satisfied with the resulting effect of the fill, make sure the transformer is turned off then you can seal off your sculpture at the constriction using a hand torch. When cool, your sculpture will be ready to display.

Now the filling process is complete, make sure the gas bottles are sealed off and all the stopcocks are closed. The manifold can now be vented. Only now is the pump turned off. Turning off the pump before having released the vacuum will suck the pump oil into the manifold — you really do not want to be doing that!

Gas mixture recipes: Single electrode plasma fill combinations

Basics:

- Some combinations will need wall contact to form filaments
- A lower-pressure fill = Glow
- A higher-pressure fill = Glow and ark
- Krypton or Xenon needed if plasma discharge is required

Neon in clear glass: Pink glow, no plasma

Xenon and Nitrogen: Green and plasma filaments

Neon and Xenon: Violet/pink with white plasma filaments

Argon: Just white filaments

Krypton in uranium glass: Green glow and plasma

More adventurous combinations:

Xenon doped with iodine (20 milli Tor): Electric blue glow with white and blue filaments.

Xenon doped with Ethanol (C_2H_5OH): Green glow with cool dancing filaments.

Neon (470 Tor) doped with Krypton (1.4 Tor) and Oxygen (1.1 Tor): Blue and Red glow with pink filaments.

Neon (345 Tor) doped with Nitrogen (10 Tor) and Krypton (10 Tor): Golden orange.

Krypton (100 Tor) or Neon (450 Tor) doped with Iodine vapor: Gives a blue hue with filaments.

Even simpler

Even simpler than one electrode is no electrodes, none — completely electrode-free!

The procedure for filling is remarkably similar, however, as there is no direct electrical connection to the glass, the gas has to be excited by means of a nearby transformer. The same "driver" can be used, but instead to connect to an electrode it must be built into a kind of radio transformer base.[1] The filled sculpture can then be placed upon the base to be energized and displayed. With this technique you are limited to relatively small sculptures.

A transformer base would be fairly simple to construct, however, as I am no electronics expert, I would not begin to offer advice on design and construction but would suggest you speak to someone who is.

Gas mixture for electrode-free fills

If plasma discharge is required, you will need to fill using Krypton or Xenon.

---

[1] *Electrode-free plasma sculptures can be energized in a microwave.* Hardware: Neon21 Driver — can be purchased from *Information Unlimited NH USA.*

Xenon (80 Tor) top pressure — mostly white with dancing plasma, can be difficult to arc.

40–80 tor is very satisfactory — more active filamentary discharge.

Less than 20 Tor not at all good! — Dim diffused emerald.

Xenon (10 Tor), Iodine (20 milli Tor) and the balance made up of other noble gases to a total pressure of 320 Tor.

A significant financial outlay is required to purchase the necessary equipment and materials to pump and fill a sculpture. There are professionals out there, particularly in the neon industry, which could be employed to do this for you.

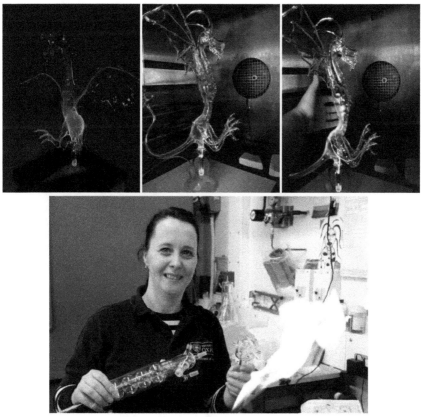

Mrs Terri Adams FBSSG Senior Scientific Glassblower, Oxford University.

Chapter 13

Sealing in Frits in the Frosty North

Philip Legge

Scientific Glass Design Ltd., ON, Canada

What is a frit? Technically, frit is a small granule of glass. The granules of borosilicate or quartz glass can be packed and baked to just above the melting point to gently fuse them together. This makes a porous block that can be sliced and shaped to form disks. These disks can then be used to filter liquids or finely disperse gases into liquids. A finer or smaller frit makes the porosity of the disk tighter or finer while a larger frit creates disks with more open or courser porosities. In North America we call these disks made of frit "fritted disks", or "frits" for short. Our friends overseas tend to call them "sinters". I am not too sure where that term (sinters) comes from, but for the purposes of this chapter I will refer to them as "frits".

The fritted disks are commonly sealed into glass tubing (usually on a glass lathe) to make filtering funnels, liquid chromatography columns and gas dispersion tubes. They are quite common in most glass shops, but many glassblowers dislike working with them. Too many glassblowers tend to regard frits as problematic, difficult, finicky or perhaps some other type of salty language, but why? Well frits are fuzzed to the inside of a tube of Pyrex (sounds simple right?), but because they have a large amount of air in-between the fused granules, they conduct heat very differently than solid borosilicate tubing. The air within the disk insulates, causing the frit to heat and cool much more slowly than the tubing which surrounds it, so when a heat source is suddenly added or removed, the glass surrounding it tends to expand or contract quicker than the disk. This causes a catastrophic crack either in the tube or the frit itself, which is difficult or sometimes impossible to repair, especially in larger diameters.

When I was an apprentice, I was given all kinds of advice by fellow members of the ASGS (American Scientific Glassblowers Society) on how to work with frits. I had the opportunity during conferences to try the varying techniques and tricks that my instructors had developed over the years. They all worked to a varying degree, but in my opinion, they were all far more complicated than need be. One method was to craft some holders that would be wrapped in fire retardant tape, which supported the disk within the tube, preventing it from moving. It worked well, but if you were not incredibly careful, the tape would become hot, off-gas and foul your work with a gray black film. Also, after a successful seal was made you had to keep the disk hot while you removed not one but two holders. That was the point where my piece cracked. Another way I was shown was to hold the disk with a vacuum jig in the lathe's tailstock. That way worked well! I liked it, but the drawback was it takes me 20 minutes to just get the vacuum jig and vacuum pump hooked up and if you have to put a stopcock, joint or valve on the bottom of your piece, you have to disassemble the set up while keeping the frit hot so it doesn't crack. Then there were the methods of doing them freehand on the bench... I am not even going to bother explaining those, they were all needlessly complicated and my brain quickly deemed those methods to be not worth the storage space.

Every glassblower seems to have an opinion on how to do them and why they break. I've been told dozens of times when I explain my method, "I can't do them that way, it's too humid in the south! Frits hold moisture and will crack" or "The temperature in Chicago varies too much and they (the fritted disks) will get shocked". Many say to anneal them first, or warm them in a toaster oven, store them in Tupperware containers with humidity control packs, or don't bother doing them because you can buy fritted columns already completed. Well, I live just east of Toronto, Ontario, Canada, and work in a town called Ajax. The glass shop is located just a kilometer north of Lake Ontario, one of the largest freshwater lakes in the world. We have cold winters, lake effect snow, ice storms, hot summers that are both extremely humid and sometimes dry, as well as wet springs and falls. Temperatures range from $-25°C$ to $38°C$. The weather has never affected my success rate in working with frits.

Now I call this my method, I have used it for nearly 20 years, so it feels like mine, but it was taught to me by my father who is also a glassblower. He always figured out ways of doing things that are both cost-effective, produce consistent results and are quick. You cannot spend half a day making one chromatography column or fritted Buchner funnel when you run your own business. Last week I had an order for two fritted columns, one 150×700 mm and one 125×180 mm, both with a coarse frit, high vacuum valves and 45/50 outer joints on top. I started Tuesday when i got into work and had both completed just minutes before lunch break. I do not say this to brag, but only to show how quick and easy the work can be done.

The smaller column will be described first. This method works well from diameters of 15–125 mm. You will have to use your best judgment of which torches to use. (You most likely would not need a carriage burner for anything under 80 mm diameter). Turn on your oven, do this right away. It should be at an absolute minimum of $500°C$ (though 540 degrees is better especially for the larger fritted disks) by the time you have sealed in your frit. I began by placing a length of 125×3 mm Simax tubing in the lathe and making a scratch with diamond tool at 300 mm in length. At the mark I pulled down the tubing to make a round bottom and affixed the 45/50 joint and the high vacuum stopcock. I flame annealed the work I had just done and allowed it to cool for a few minutes. Once cool enough to handle, I measure 180 mm of straight tubing and make a mark with my diamond tool. This mark is where the frit will be sealed in. Put your piece back in the lathe with the round bottom in the headstock and the open end in the tailstock. At the mark, heat a thin strip of the tube with the carriage burner and use a graphite tool to make an indent (Figure 1).

Figure 1.   Use of graphite tool.

This indent becomes a seat for you later to press the frit up against. I was using a 110 mm frit and the inner diameter of the tube was 119 mm, so I had to tool the indent to be 5–6 mm deep. Stop the lathe and put on a pair of gloves to protect your hands from the heat of the tube and put the cold/room temperature frit inside the tube, as close to the indentation as you can get. A scrap piece of tubing with a very flat end is placed in the tailstock and used to butt the frit up against the indent or seat (Figure 2).

It is important to get the frit in place while that indent is still extremely hot. If you have your materials and tools handy, you should be able to get the frit in place in under a minute, which is fine. Turn the lathe on and warm up the frit with the carriage burner. The flame should have just a small amount of oxygen added and let it rotate for a minute or two (Figure 3).

After the minute is up, you can crank up the oxygen and fuse the frit to the inside of the tubing (Figure 4). A graphite tool should be used to make sure the seal is uniform around the entire disk. You can tell when a good seal has been achieved as the frit will make a solid white strip around the entire circumference. The scrap piece of tubing can then be butt-sealed to the end of the column and used to pull a round bottom (Figures 5 and 6). This is also the time to use

Figure 2.   Support of the frit.

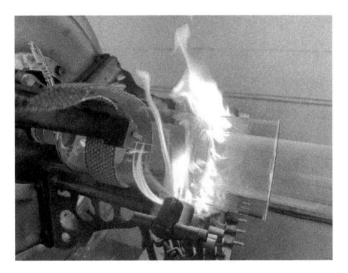

Figure 3.   Gradual heating of frit.

centrifugal force or a few good puffs on the blow hose to even out the wall thickness. Once you're happy with the appearance of the seal, the carriage burner can be set to have a gentle annealing flame and placed under the rotating frit to keep it warm while the high-vacuum

Philip Legge

Figure 4.   Sealing of the frit.

Figure 5.   Pulling of the round bottom.

valve is prepared and affixed to the bottom of the column (Figure 7). Remember the oven that you turned on a while ago? Well, shut off the fires and get it into the oven as quickly as you can. You do not need to run, but having an assistant open your chuck and the oven

Figure 6. Forming hemispherical end.

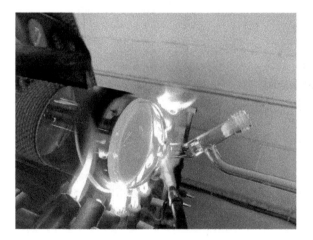

Figure 7. The valve attached.

door is a great help! I would estimate the time you have to get the piece in the oven to be about a minute. Hopefully, the oven has gotten up and above 500°C. Once the oven has gone up to annealing temperature and clicked off, allow it to cool slowly, please don't open the door too much. Once the temperature is around 200°C you're in

Figure 8.   The completed unit.

the clear and can open the door to allow your fritted apparatus to cool more quickly. At this point, sit back and admire your job well done (Figure 8). It was quick and easy, required no preheating or moisture removal and, most importantly, did not crack.

But wait, there is more! The first method works swimmingly for frits large and small, but there's another easy and quick way to do frits. The second method in my opinion is best for large frits. The smallest size I have used this method with is 110 mm and the largest is 250 mm. It's similar to the first method as you don't preheat the frit, it requires no special tools and is quick and easy. Again, the oven must be good and hot! If you are doing a frit of 150 mm or larger, please make sure the oven is as close to annealing temperature as you can get. Now for this job the column is supposed to be 150 mm in diameter. I am using 150 mm tubing with a 5 mm wall thickness cut to a length of 700 mm. It, like the first column, will have a bottom drain valve, a 45/50 joint and a high-vacuum Teflon stopcock for vacuum and inert gas. We begin this column with a 2000 ml flask

Figure 9.   The socket is removed.

Figure 10.   Tap attached to base.

blank. The flask is put in the lathe, the neck pulled off and a 45/50 is sealed on. (Figure 9).

Secondly, the bottom drain valve is attached to the bottom of the flask (Figure 10) and the vacuum take-off valve is affixed near the top. Flame anneal well so it won't have to be run through the oven. Think of the flask as a planet, the joint being the north pole and the drain valve the south pole, just north of where the equator would be is where we will make a flame cut. Using a good carriage burner with a sharp flame, proceed with the flame cut. The "northern" hemisphere will become the top of the column. Put the short length of 150 mm in the tailstock of your lathe and you can seal the top on the tube. Now 2000 ml blanks are a little larger than 150 mm, so you will have to use a graphite paddle and tool down the flask slightly, but once that is done you can seal it onto the tube and just use the speed adjustment of the lathe to control the centrifugal force to get a nice even wall (Figure 11).

Proceed to flame anneal and set the piece aside for now. The "southern" half can be put in the headstock and a piece of scrap tubing with a square end in the tailstock. Like in the first method, the scrap tubing will act as your holder. A larger size is better but not so large that it is close to the edge of the frit. (I used a piece of 95 mm.) Move the tailstock in so the scrap tubing is just outside of the flask half. Place the frit in the flask bottom and move the tailstock in gently so the tube meets the frit and holds it firmly in place within the flask half (Figure 12).

Figure 11.   Centrifugal force to produce hemisphere.

Figure 12. Frit supported.

Hopefully there is still some residual heat in the flask half, but if there is not, no worries. Turn on the lathe and with the carriage burner apply a gas-only flame for one or two minutes. After that time has passed, add some oxygen to the flame and allow it to turn in the annealing flame for another two minutes. At this point it is safe to begin sealing in the frit. Remember how this side of the flat was slightly larger? This is to prevent flame splash from getting on the frit too much. Turn up the oxygen and allow the glass to become molten around the frit. Do not apply too much heat above or below the frit because it can get a little sloppy if you do. Using a graphite rod or paddle can help the glass fuse to the frit. Like in the first method you should be looking for the same solid white line around the frit, no gap, no air pockets (Figure 13).

Once a good seal has been made around the frit, adjust the carriage burner flame to be a gentle annealing flame and leave it under the frit to keep it warm. Suddenly stop the lathe, remove the scrap tube that was your holder and put the 150mm in the tailstock and turn it (the lathe) back on. If you are comfortable with your lathe and your shop setup, this should be a quick process (10–20 seconds I would guess). Now the flask bottom with the frit can be sealed onto the column top. Like when we did the top of the column you will have to paddle down the lip of the flask a little and then it

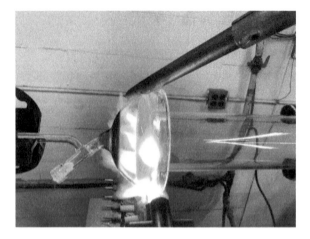

Figure 13.   White seal completed.

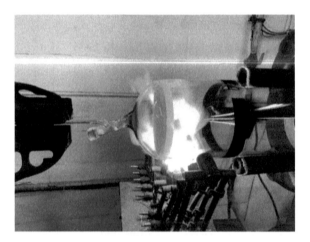

Figure 14.   Centrifugal force to complete join.

can be fused to the body, again using the speed control of your lathe to use centrifugal force to your advantage (Figure 14).

Now if you have a shop mate, I suggest enlisting their help. Have them open the lathe chuck while you remove the column and then get it into the oven quick. They can also open the oven door and close it for you. Time is of the essence, do not rush as when people rush is when mistakes happen, but be smooth, controlled and speedy.

Figure 15.   Final large diameter sinter.

Discuss where you are going and where you want your shop mate to be beforehand as you do not want to trip over each other with a large, hot column in your hands. Once the oven has completed its cycle (we anneal glass this thick at 580°C for 10 minutes with a cool down over 2.5 hours) you can remove it and admire your work (Figure 15).

This method has worked for me consistently for nearly 20 years and for my father for over 50. It is not particularly difficult. Anyone who is confident with their lathe skills can do it. Our apprentice has been doing frits this way too for a few years and you can, too. If there is any additional advice I would give, it would be to make sure all your materials and tools have been laid out close by beforehand. The odd time this method fails (which I assure you is very rare) it's due to a missing paddle, lost chuck key or accidentally grabbing two left hand gloves when you want to remove it from the lathe.

Chapter 14

Drawing on Experience: Two Methods for Holding/Manipulating Eccentric Shapes

Graham Reed

Box Shapes from Boro-glass Sheet

Without jigging the component parts and keeping hot the partially fused plates, it is exceedingly difficult to avoid stress-induced cracking. This advice also applies to larger rectangular shapes in quartz sheets.

The photos show

A drawing, one-of-two parts; bench construction; lathe joining and finished piece.

The 2 mm thick boro-sheet was water-jet cut, each of the two halves were preheated in a kiln and placed upon a hot carbon plate to facilitate the glazing of the edges. This is a vital stage and a point where the glass will crack very easily. I played a large, bushy flame over the glass before running a hot-blue flame along one edge onto which I sealed a handle at a point where later I could disguise any mark. Holding the plate by the handle (any rod or tube up to 15mm diameter) I then was able to hold the whole plate in a soft flame while using a hand-torch to glaze the edges.

The glazed, flame-annealed plate was placed upon a hot, carbon plate matching the profile [not shown], the handle was removed and then selectively heated to fold the sides that later completed the box.

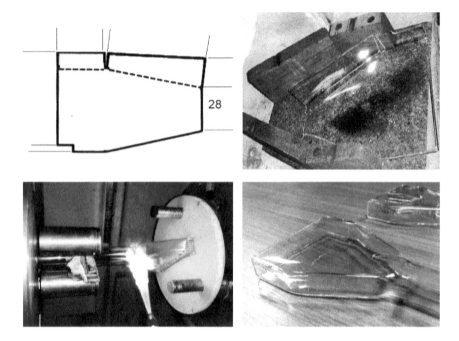

All the way through these stages, a bushy flame was washed over the work piece.

The two mirror halves were mounted in the carbon jig [shown] with a spacer inside to support the two parts before sealing along the edges. It cannot be over-emphasized how vital it is to keep everything hot during the sealing phase and having a hot oven close by to return the completed box. The jig was supported so that a "Bunsen" flame heated it from below.

The finished job has a 30 mm dia. tube attached. The solution to supporting the box was the lathe jig. It consists of a 10 mm thick plate made from GRP (glass-reinforced plastic — yachting offcut!). I cut a circle to avoid having corners spinning in the lathe. The crucial part of the jig is a slit running to the circumference from the rectangular slot so that the opening/closing of the jig using the lathe jaws allows the slot to grip the box. I decided not to blow the glass joint between the box and cylinder but was able to insert a carbon pencil on a long holder to tool the joint to the correct bore.

Always try to avoid making a jig that only has one product, so I experimented with it and found that I can hold any object — cone or socket joint, screw fittings, etc. so long as a part of it has a size matching the slot. It could easily hold a U-bend to allow adding a tube to convert it into a Y-shape: magic!!

Large-Scale "Semi-Tech" Plant Glassware

The late and much missed Fred Morse in his Southampton factory was asked by a large "Pharma" customer to add a 50 mm QVF® outlet on the side of a 450 mm dia. pipe section of 600 mm length. Many a glassblower might approach this challenge by rotating the main body in the same way as one will hold any cylinder. Heating such a large diameter body and then spot heating it to pull a hole ready to accept the sidearm is both risky and adding to the difficulty of making a clean fuse between the main body with a wall 10–12 mm thick and the sidearm with a 3 mm wall.

The jig solution required a large holder to enable the body to be "tumbled". A –70 mm steel pipe was v-notch cut at one end to hold a 75 mm wide angle-iron that was welded into the notch. At each end of the angle strut, a 13 mm hole was drilled. A steel bar 10–12 mm diameter was made of the same length as the angle bar. Two long bolts of 10–12 dia. were modified by cutting off the head and hot-fashioning a loop on the plain shank that hooked onto the bar.

The manufacturing process

Hold the jig in the headstock of a lathe with a large swing — 400 mm or more! Have an assistant lift the 450 dia. tube up to lay along the angle slot while the steel bar is placed inside the pipe. The two looped bolts are to be set into the end holes with their washer and nut set ready to accept the bar. When tightened, the bar will hold the tube in place; fireproof packing along the bar will help grip the job and avoid scratching the inner wall of it.

The tailstock will have the sidearm piece set ready to make the seal. The lathe speed must be kept low to avoid centrifugal stresses;

there is a lot a weight swinging! Heat the body at the centre of the
rotating glass and pull a hole in the wall and shape it to the matching
diameter of the sidearm. Fused in the normal way, a carbon rod on
a long handle can go through the tailstock to shape the joint. A
"good" soak with large burners around the joint is essential before
removing the hot job from the lathe and into a preheated oven. Plenty
of heatproof protection is essential clothing! Without any "photos"
to show the rig in use, I have made these sketches to illustrate the
salient features.

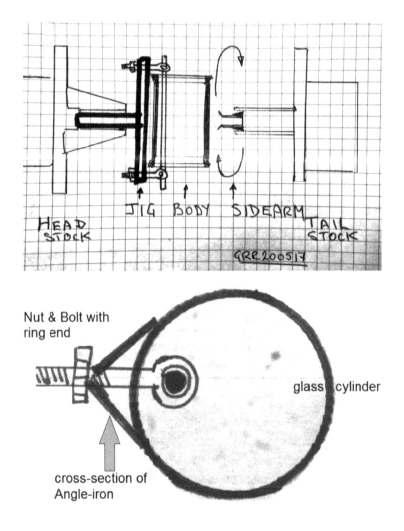

Chapter 15

Reinhold Burger X-Ray, Dewar or the Thermos

Norbert Zielinski

Scientific Glassblower, Technische Universitat Berlin, Berlin

The first glass instrument factory was founded in 1894 in Berlin before the beginning of the first world war and it had around 40 employees.

Independence Learned in America

Reinhold Burger completed his apprenticeship at the age of fifteen with the company C.A.F. Geissler & Son in Berlin. He was employed as a glass technician at Siemens & Halske. The prosperous economic relations also between Siemens and the United States before the First World War encouraged Burger to deepen his knowledge of physics and manufacturing during two visits to the US. The main reason for his first trip was to find and build a thermometer factory with a business partner in the USA. On his second trip to the United States, he took a large variety of equipment and tools with him as he intended to settle down in Philadelphia. Despite limited initial capital, he managed to make a success of the thermometer company. Reinhold Burger returned to Berlin after two years. He joined a partnership for delivering training but as well founded his own company in Berlin at Chausseestrasse 2E in 1894. This company, which was founded in the German Empire, survived the First World War, experienced the Weimar Republic and the Third Reich, the Second World War and continued to produce in the GDR for many years. In the beginning, he initially produced thermometers, then mainly manufactured X-ray tubes, glass vacuum systems and produced the first X-ray tube in Berlin. He presented Conrad Röntgen's discovery a few days after the publication to an expert audience in the Urania Scientific Institute in Berlin in January 1896. He immediately had many customers within the Berlin research chemists.

The Thermos Bottle Patent

Quote from an interview from 1941:

> "The invention of air liquefaction gave me the opportunity to further improve the vacuum flask for minimising heat loss. In the absence of liquid air, I checked the usability of the containers with

hot water, since most existing containers did not meet suitable standards. It occurred to me that they could be used for drinks instead. I picked up a series of small spherical jars I had made earlier and filled them with hot coffee, tea, milk, and the like. After 24 hours, the drinks were as ready to use, as if they had just been prepared ...".

The Munich-born inventor and ice machine manufacturer Carl Linde commissioned Reinhold Burger for the development and manufacture of glass insulated containers to store liquefied air at $-194.5°C$ to stay available for science experiments if possible.

The original idea for an insulating container made of glass was created by the Scottish physicist Sir James Dewar when he developed an insulating vessel for physical laboratory tests in 1874. The aim was to store liquefied gases at low temperatures. The first vessel was made of two glass tubes, one inside the other, sealed at the top. He pumped the air out between the glass tubes. However, they were completely unsuitable for storing liquid air. Reinhold Burger adopted the principle of Dewar's vacuum flasks, redesigned them and found a good use for the masses. It is an ongoing story in the Burger family how around 1898 Reinhold Burger tested his insulating vessel with hot water instead of expensive liquid nitrogen to see how long it kept its temperature. He realized that it would be great for tea and coffee as well. So he created a catchy rhyming advertising campaign: Keeps cold and hot without fire — without ice, in German **hält kalt und heiss ohne Feuer — ohne Eis.**

In many of the first double-walled glass containers, the inner vessel broke away from the upper neck base due to the weight of the internal liquid. Since Reinhold Burger only had very simple types of glass available at this point, he made many experiments to find

Patented and protected in all Unmissable for a member statescultured person.

the right measure of glass thickness for stability. Thick-walled necks were quick to crack when they came into contact with hot or cold liquids. In order to achieve a mechanical relief of the neck during pouring, small asbestos supports were placed in the lower area of the inner container. Initially the silver plating was not stable enough and peeled off after some time. He had to make many attempts to find a recipe to create a durable and long-lasting type of silver plating. In 1900, he finally managed to patent a vacuum flask for everyday use with a sturdy metal sheet jacket. The patent has the

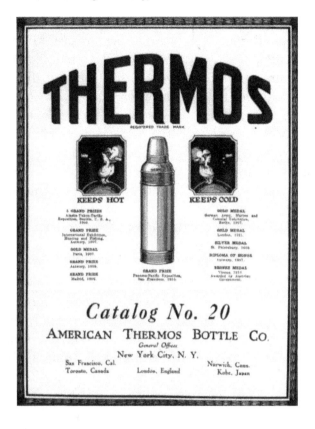

number DRPNR. 170057 of the Imperial Patent Office and was issued on October 1st in 1903. In the following year 1904, he had his creation "Thermos" protected as a trademark. In 1906, the patents were registered in the USA, England and France and in many other states. He received a gold medal at the World Expo in St. Louis, USA. In 1904 and in Milan, Italy, in 1906 he received another gold medal and an honorary diploma. In 1907 he was honored at the Grand Prix of the hunting and fishing exhibition in Antwerp and also received another gold medal of the German Army, Navy and Colonial Exhibition in Berlin in 1907. In 1909, the thermos bottle was awarded the Grand Prize at the Alaska-Yukon-Pacific Exposition at the Seattle World's Fair. In 1911, he received a silver medal at the International Hygiene Exhibition in Dresden.

In 1906, Reinhold Burger founded the Thermos-Gesellschaft m.b.H. in West Berlin. The advertisement reads: "Thermos keeps drinks cold without effort and without chemicals, hot drinks stay hot for 24 hours, cold drinks, even on hot summer days, stay cold for days without ice. Indispensable for tourists, travellers, drivers, cyclists, water sports, the military, airships, forestry officials, hunters, offices, factory workers and all other workers as well." In the first few years, sales did not go as well as expected given the huge amount of advertising placed in the relevant daily newspapers. The department stores only took thermos bottles on commission. Reinhold was more of a developer and inventor than a businessman. Therefore, in 1909 he sold this business area to the Berlin Charlottenburger Thermos Aktiengesellschaft, which he had founded. Thanks to an impressive sales volume of 495 shares of 1000 Reichsmark each, Burger received so much money from his shares within his Thermos-Gesellschaft m.b.H. that he was proud to become the second largest taxpayer right after the cigarette manufacturer Mr. Garbaty in Berlin Pankow. In 1909, he transferred all foreign rights to the American Thermos Bottle Company in New York, to Thermos Limited (Tottenham) England and to the Canadian Thermos Bottle Ld. The worldwide right to brand name "Thermos" was part of this deal. The largest thermos factory in the world at that time was built in Brooklyn, New York. With his wealth of experience Reinhold Burger led setting up the manufacturing process in the United States from the beginning. The American factory workers called him "professor". During the construction of the production line, R. Burger ordered many machines and devices from Germany and put them into operation in New York, Brooklyn, in 1907. Due to particularly long working days during the beginning of industrialization in the USA, the thermos bottle from Mr. Burger from Berlin began its triumphal success in the States and then conquered the rest of the world. Due to the general economic decline after the First World War, the Burger family had been economically worse off from year to year. It is still reported from a memorable visit in 1921 by the director W. Walker of the American Thermos Bottle Co., who invited the Burger family to the Hotel Adlon in Berlin for dinner. The burger sons continued to recount the lavish ambience and fantastic food for many years, while everyday life at that time only offered cabbage soup. Mr. Walker was now a dollar millionaire, while Reinhold Burger almost lost his economic possessions due to the First World War. Since demand and therefore production fell to a minimum, the business, which initially had 20 employees, was reduced to a family business in 1926 and moved to the residential building at Wilhelm-Kuhr-Strasse 3/Pankow. Reinhold Burger worked here until his death in 1954. One of the sons operated this company until 1982 under the known difficult conditions in the GDR (e.g. the required oxygen was obtained from a neighboring locksmith's shop). Although this company was a recognized supplier to the research sector in the old federal states, the official bodies kept putting obstacles in its way. A successful craft business did not seem to fit into the official GDR worldview.

Packing label USA.

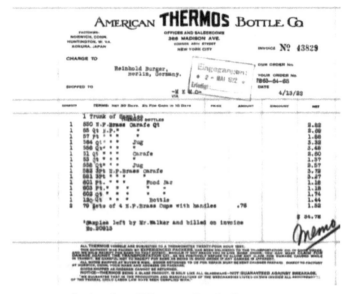

An invoice of Thermos Bottle Company.

The American Thermos Bottle factory in Norwich Connecticut, 1925.

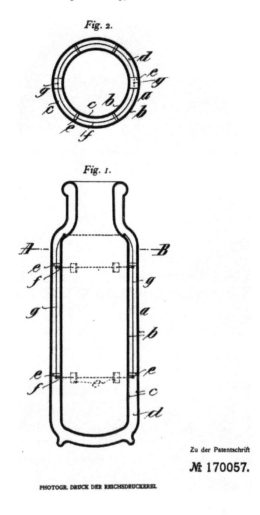

Fig. 2.
d
e
g
a
b
c
b
g
c
e
f
Fig. 1.
A
B
e
f
g
g
a
b
e
f
e
c
d
Zu der Patentschrift
№ 170057.
PHOTOGR. DRUCK DER REICHSDRUCKEREI

Imperial Patent Office

Patent application #170057 Class 64a.

Double-walled vessel with the air evacuated between the two bottle walls.

It is known there are glass vessels with double walls and an evacuated air-free space which has a lining made from an elastic material that poorly conducts heats and stiffens the inner vessel to prevent rupturing if transported when filled with liquid.

Reinhold Burger in Berlin, Patented in the German Reich on the 1st of October 1903.

The invention here presented is the small asbestos particles or another poor heat conductive material attached to thin wires then added to the lining placed close to the bottom of the vessel.

The lining is made of a thin ring-shaped wire, which is appropriately tensioned to ensure that the attached insulating material does not move out of place. To prevent the inside of the glass walls, which are covered in a mirror-like substance, from being damaged, and accordingly, the properties of the glass vessel to prevent temperature changes within would be greatly diminished.

On the accompanying drawing is a possible example of the invention.

Figure 1 is a vertical section through a vessel according to Sir James Dewar.

Figure 2 is a section online A-B in Figure 1.

The vessel has double walls (b) which are provided with a mirror coating (c)

In the enclosed wall space (b) is an empty cavity (d) which has a lining (e) made of asbestos cardboard paste or the like that touches the walls (b) expediently or only has a tiny distance to them.

The lining (e) is attached in groups on thin wire (f) and these are suspended on more wires (g) that are held together by a ring-shaped wire. The housing is designed for use as a domestic utility object and intended as a bottle for combat.

Patent claims:

1. Vessel with double wall with an evacuated air cavity enclosed by glass walls with an effective lining that stiffens the inner vessel, attached with tiny plates or similar to a ring-shaped wire frame.
2. The shape of the vessel according to claim 1 is predominantly characterized by a wire holding the plates (e) itself attached to a ring-shaped wire frame (f) where (g) is suspended.

Science and Medicine

Reinhold Burger 1951 with his grandson Axel Burger who can be seen here actively helping his research.

Axel Burger and Norbert Zielinski.

KAISERLICHES   PATENTAMT.

AUSGEGEBEN DEN 22. APRIL 1902.

PATENTSCHRIFT

— № 129974 —

KLASSE 21g.

REINHOLD BURGER in BERLIN.

Vorrichtung zur Erzeugung von Röntgenstrahlen.

Patentirt im Deutschen Reiche vom 19. April 1901 ab.

Die bei Erzeugung von Röntgenstrahlen eintretende Erwärmung der Antikathode ist ein großer Nachtheil, weil erstens die Antikathode beim Glühen leicht zerstört wird und zweitens das Vacuum in der Röhre sich ändert, wodurch die Röntgenröhren bald unbrauchbar werden.

Es sind nun zur Beseitigung dieser Uebelstände von verschiedenen Seiten Anordnungen getroffen worden, um die Erhitzung der Antikathode durch Wasserkühlung zu vermindern. Jedoch erreichen derartige Ausführungen sehr unvollkommen ihren Zweck, weil das zu kühlende Metall der Antikathode entweder nicht direct in die Kühlflüssigkeit eingeführt wird, oder sehr unvortheilhaft construirt ist, wie z. B. bei der in »Rayons cathodiques et Rayons X« von J. B. Breton (S. 90, Fig. 124) beschriebenen Röhre. Die Kühlflüssigkeit ist hier wohl direct mit der Antikathode in Berührung und dürfte auch gut kühlen, aber ein dauernder Gebrauch dieser Röhre ist ausgeschlossen, weil die Metallröhre eingekittet ist, und infolge dessen nicht luftdicht abschließen kann.

Die ähnlich construirte Röhre, welche in der Monatsschrift »Fortschritte auf dem Gebiete der Röntgenstrahlen« auf S. 224, Fig. 2, beschrieben ist, hat schon den Vortheil vor der vorerwähnten Röhre voraus, daß die Verbindung zwischen Glas und Antikathode ein auf beiden Seiten offenes in das Glas luftdicht eingeschmolzenes Platinrohr bildet. Die Verlängerung der Platinröhre besteht aber aus einem angelötheten Metalltrichter, an dessen breitem Ende die Antikathodenplatte ebenfalls

aufgelöthet ist, ein Fehler, welcher erstens die Röhre der Gefahr des Ablöthens der Platte aussetzt und zweitens ein dauerndes Vacuum, wie Röntgenröhren es erfordern, ausschließt.

Eine andere wassergekühlte Röntgenröhre ist die in der deutschen Patentschrift 119507 beschriebene. Hier ist die Antikathode über einer Glasröhre befestigt, in welcher sich die Kühlflüssigkeit befindet, die aber nicht mit dem Metall in directe Berührung kommt; eine energische Kühlung findet deshalb nicht statt.

Die luftdichte Einschmelzung eines massiven starken Metallstabes in einer Glasröhre, welcher in die Kühlflüssigkeit führt, wäre wohl das Vollkommenste, ist aber wegen der verschiedenen Ausdehnung des Metalles und Glases nicht haltbar.

Die vorliegende Construction hilft nun diesem Uebelstand in günstiger und einfacher Weise ab. Die Antikathode ist an einer Platinröhre befestigt, in welche ein Metallstab oder Rohr (am besten Kupfer) eingesteckt ist, die direct nach außen in die Kühlflüssigkeit führen. Die Einschmelzung von Platinröhren in Glasröhren ist wohl schon öfter ausgeführt, aber nicht in dieser zuverlässigen Weise mit einem geschlossenen Ende in Anwendung gebracht worden.

Die Erfindung ist auf beiliegender Zeichnung im Durchschnitt dargestellt.

In dem äußeren Glascylinder *a* hängt die vom Vacuum umgebene Glasröhre *b*. Mit letzterer ist das an einem Ende geschlossene Platinrohr *c* luftdicht verschmolzen. In dem

Imperial Patent Office

Accepted on the 22nd April 1902 Patent application #129974 Class 21g.

Reinhold Burger in Berlin
Device to create X-ray beams.

The process of heating the anticathode during the production of X-ray beams has a great disadvantage because firstly the anticathode is easily destroyed when glowing and secondly the vacuum in the tube changes, which deteriorates the X-ray tubes rapidly.

Various suggestions have been made to eliminate these issues by preventing the heating up of the anticathode by water cooling. However, this design does not achieve great results because the metal of the anticathode, which needs to be cooled, is either not directly submerged in the cooling liquid or it is constructed unfavorably as for example in the "Rayons cathodiques et Rayons X" by J.B.Breton (p. 90, Figure 124) described tube. The coolant is in direct contact with the anticathode here and should cool well, but permanent use of these tubes is limited due to the metal tube being bonded and as a result it cannot be hermetically sealed.

A similarly constructed tube — featured in a monthly publication "Advances in the field of the X-rays" on p. 224, Figure 2 — has an advantage over the previously mentioned tube, because the bond between the glass and the anticathode is made of fully sealed melted platinum tube. The extension of the platinum tube is made of a soldered metal funnel with an anti-cathode plate attached to the wider end, a mistake which firstly exposes the tube to the danger of losing the soldered connection and secondly it cuts out the option of having a consistent vacuum in the X-ray tube.

There is another water-cooled X-ray tube described in the German patent specification 119307.

Here the anticathode is attached to the glass tube which leads into the cooling liquid but coming into direct contact with the metal therefore dynamic cooling does not take place.

A solid strong metal rod sealed into a glass tube leading into the coolant liquid would be the most ideal sounding idea but is untenable due to different expansion rates in metal and glass.

The following construction helps to prevent this deplorable situation in a cheap and simple way.

The anticathode is attached to a platinum tube, which is covered with a metal rod or tube (preferably copper) that leads directly outward into the coolant liquid. The process of melting platinum into glass tubes has been achieved several times, but the closed end was not reliable.

The invention is shown in the accompanying drawing. In the outer glass cylinder (a) hangs a glass tube surrounded by vacuum (b). The latter is closed at one end and fused airtight to the platinum tube (c).

The "cold red light radiation apparatus" was patented in 1927 in collaboration with doctors of Berlin's Charite university followed by many other medical therapy and diagnostic devices.

Axel Burger reports about an order from the Kremlin Hospital for a well-made German medical device, a cold light irradiation device, obviously intended for Stalin in the early 1930s.

This regular order — reviewed by the Russian officers after the end of the war — became substantial for the survival of his small craft business over the next few years, since Berlin and the Pankow district were under Russian control.

At the turn of the century, Reinhold Burger successfully patented and manufactured an improved X-ray tube, German Reichs Patent No. 129974 from 1901.

This patent was inspired by a collaboration with Conrad Röntgen's employees — the first Nobel Prize winner in physics. One of the original X-ray tubes with which Mr. Röntgen personally experimented can still be seen in the Burger Museum in Glashütte / Baruth. The X-ray tubes manufactured in various designs from 1900 are characterized by a protective layer containing lead. An important point was the significant reduction in heat generation. The first serial production for X-ray tubes was started in Berlin at the Reinhold Burger company. As a result, a wide application in medicine followed. They were represented in England and France.

When looking through the correspondence, we found letters and orders from all over the world: In October 1933, 36 X-ray tubes were delivered to the Friedländer company in Chicago. We also found orders from New York in 1937, from Berlin in 1902 from the Institute for Physics at the Charlottenburg University of Technology, from Berlin in 1901 from Siemens & Halske, from Bombay in 1936. Conrad Röntgen received the first Nobel Prize in Physics in 1901 for the development of the X-ray tube. The rightful inventor Reinhold Burger had to fight several patent litigation cases but repeatedly filed more patents in various areas, such as a patent for a throwing game for children or a spark plug tester for the automotive sector.

REINHOLD BURGER in BERLIN.

Vorrichtung zur Erzeugung von Röntgenstrahlen.

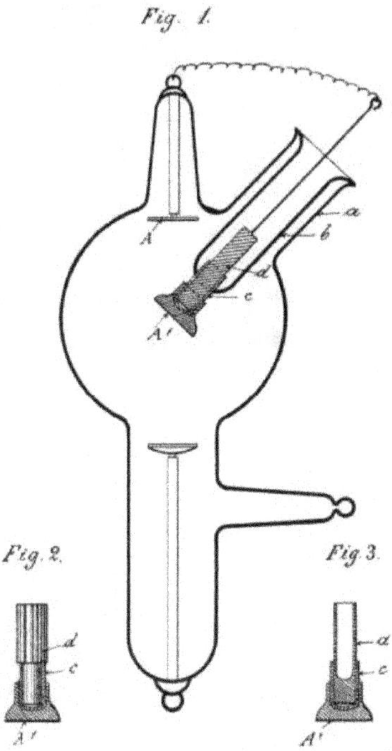

Fig. 1.

Fig. 2.

Fig 3.

Zu der Patentschrift

№ 129974.

PHOTOGR. DRUCK DER REICHSDRUCKEREI

H. Burger in 1954 in front of the showcase, which can still be seen in the museum in Glashütte/Baruth.

The city of Berlin named the Reinhold Burger high school in his honor and at WilhelmKuhr-Str. 3 in Berlin Pankow there is a plaque with the inscription:

> "Reinhold Burger Scientific Glassblower and holder of numerous patents — born 12th January 1866 — died 12th December 1954 — lived and worked in this house from 1927 until his death. He invented the thermos bottle in 1903".

The entrance to Wilhelm-Kuhr-Strasse 3

I would like to thank Axel Burger for his support, without him, this chapter would not have been possible. Many thanks to Heinz Krohn and Günter Palm for the film and photo shoots.

N. Zielinski translated by Stephanie Kubler Preston

Chapter 16

Pull the Glass with the Flame

Jens Koster

Chemistry Department, University of Hamburg, Hamburg, Germany

In order to "pull" glass with a flame, it is important to understand the principle behind it. Glass in its liquid state is subject to certain physical laws that play an essential role here. On the one hand is viscosity, which is dependent on the temperature, and on the other hand, is adhesion, which in turn depends on the viscosity and thus also on the temperature. As a memento, the higher the temperature, the more and faster the glass wants to flow to the hot spot. What that leads to, especially at the seams, is that the glass collects. If you want to get the glass back into a homogeneous wall thickness, you can inflate it and pull it accordingly. That is how it is generally taught and that is how I have learned it, too. But there is another way to get the glass homogeneous again without having to pull or do anything else. Which is not even possible with certain systems if you, for example, think of steam pipes that are melted onto a solid body. In order to distribute the glass accordingly, one can make good use of the adhesion and make use of the above-mentioned relationship.

You heat the thick area until it glows slightly red and then pull the glass with the flame to the too thin area by causing the glass to run there through the heat. The seam can be moved back and forth in the flame in order to finally keep the entire width evenly warm. If the same temperature prevails everywhere and therefore the same adhesion, the glass cannot help but distribute itself evenly.

But remember, the colder the glass, the longer you must wait, the hotter the glass, the more it will accumulate from the cold edge areas. It is therefore advisable to warm up the transition to the cold area slightly and not let the glass get too hot. A slightly red sheen (800–900°C) has proven itself from my experience. The flame build-up on the burner can also be used here. Behind the oxygen cone is the hottest point at over 3000°C, which is admittedly exceedingly small. The flame becomes correspondingly colder toward the edge area, which depends heavily on the burner setting and type of gas. You should therefore try it out yourself on your favorite burner. Which, if you apply this principle, makes an additional hand torch superfluous for small diameters and normal-walled tubes. Which ultimately only leads to the need to redistribute these glass collections afterward. If you already use or must use a hand torch, you should make sure that as small an area as possible is heated. This prevents too much additional glass from running into the seam from the wall.

Have fun trying.

Chapter 17

The Original Ebulliometer and the Discovery of Polythene

Paul Le Pinnet

Fellow of the British Society of Scientific Glassblowers

In 1966, at the age of 20 my first seven years' training as a Scientific Glassblower was at Manchester University. That special day came when the man who was training me (David Greenhalgh) told me to go to the Mass Spectrometer room where the M.S.2 (the second spectrometer ever made) required some work. I went and saw that a line was required from a 1 liter gas preparation vessel which was attached to the side of the machine to the source inlet. I took various measurements and returned to the workshop, then once more returned to the mass spec room with all the equipment and materials that I thought I might need. After I had finished, the technicians, Doctors and Scientists who had all been watching said that they would have to evacuate the system to make sure that there were no leaks. In my head I had a little tantrum, "Leaks! What do they mean Leaks?" I thought to myself. Although I immediately calmed down when they offered me a cup of tea while we waited for the system to evacuate.

To my surprise and delight the tea came in a china cup and saucer with "roses round the rim", the kind of thing that ladies of a certain age love to drink from together with two Cadbury's Chocolate Digestive Biscuits. This made a deep and lasting impression on me. I decided there and then that I would always endeavor to keep my work standards as high as I possible could if this were to be the reward.

In 1973, after seven years at the University I moved to the ICI Organics Research Centre at Blackley Manchester. The Company had been formed in 1927 by the amalgamation of Brunner Mond (salt chemistry), Alfred Nobel (explosives), British Dyestuffs and United Alkali. Apparently, the first thing the new company did on formation was to establish Research Centers in each of its divisions. One of these was at Winnington Cheshire. The Head of Research was Francis A. Freeth, he realized that it was so important to have good quality technical support for his Scientists.

Freeth traveled to the University of Leiden in Holland and saw their training schools for Scientific Glassblowers and instrument makers.

Freeth was there for four weeks and eventually managed to entice one glassblower and two instrument makers to join him at Winnington, not only to provide technical support but also to pass on their skills to others. A glassblowing workshop at Winnington was built and equipped to the Leiden design and standard. In later years all if not most *ICI glassblowing workshops were constructed and equipped to the Leiden standard.*

At Winnington there was also a high-pressure laboratory which was used to understand how chemicals reacted at 3,000 atmosphere pressure. One evening someone failed to switch off one of the pressure vessels as they left at the end of the day! Twelve hours later when the vessel was opened the chemicals within had altered beyond recognition. The glassblower and instrument makers were asked to design and make an "ebulliometer" which was then used to find the melting point of the unknown material. This proved to be a polymer of ethylene, the date being 1930.

The Scientists attempted to reproduce this new polymer without success until it was found that the original raw materials first supplied to them had been contaminated and gave off excess Oxygen during the reaction and that the excess Oxygen acted as a catalyst in producing the Polymer.

Polymerized Ethylene or "polythene" proved to be an exceptional electrical insulator, far better than what was available at the time, which was: natural rubber, Guta Percha which is the juice of the Malayan gutta percha tree, which was used for insulators and shellac a thermo-plastic resin obtained by purifying the resinous excreta of certain jungle insects.

A year later Doctor John Crawford working in Ardeer Scotland discovered Methyl Meth Acrylate, which was later called PERSPEX. This clear light acrylic could be formed into complex shapes and being far lighter than glass was ideally suited as a bubble canopy giving the pilot an unrestricted view. The first planes fitted with such a canopy was the Spitfire and the P51 Mustang, modern aircraft are still fitted with the Acrylic canopy (see Figure 1).

Eventually both Perspex and Polythene were then developed in tandem becoming the I.C.I. Plastics Division.

Some years later Sir Peter Allen, Chairman of I.C.I. Ltd, said and I quote:

"Without the help of these Hollanders it is by no means certain that Polythene would ever have been discovered".

What an incredible accolade for a Scientific Glassblower and two instrument makers.

Figure 1. Bubble canopy.

These men were always referred to as Hollanders because throughout the 1930s the situation in Europe had begun to deteriorate, it was thought best not to use their proper names as they still had relatives living in Holland.

In 1980, after seven years at the Organics division I was "redeployed" from Blackley, to Mond Division Winnington Cheshire, and while there I worked with Mostyn Hart who was the British Society of Scientific Glassblowers' Board of Examiners competition secretary, he was forever telling me that he had people entering the David Flack artistic competition and that he had a full quota of trainees entering the various competence level examinations, but was concerned that he had very few entries for the Norman Collins memorial trophy which was for established and experienced Scientific Glassblowers. Mostyn was convinced that more qualified and experienced people should Enter! Eventually, I got the message. Fortunately, I became aware that Mostyn was due to go on a weeks' holiday and would be leaving the following Friday lunchtime. Once I was sure that he had left the site, I looked through the stock of old drawings but could not find anything I considered suitable to enter the competition, I went into the large store room and after a great deal of searching eventually found a box covered in dust, on which there was a label which read Ebulliometer 1930! I opened the box and found that 2/3rds of the glassware was intact, but the bottom section had degraded to dust, fortunately there was also a working drawing in pencil on brown paper. I measured what was left to confirm the dimensions (internal and external). I knew instantly that this would be my challenge and that I had a week before Mostyn returned (Figure 2).

I could see that the inner shroud within the main body was critical as once in place it would need to be re-heated to annealing temperatures (560° centigrade) probably three or four times as the work progressed. I set about attaching six stout rods to the inner shroud at the bench and adjusted their lengths, then supported the shroud within a suitable piece of tubing from which to make the inner body, I supported the two items in the lathe, I was then able to "run in" (i.e. Melt) the rod attachments with a

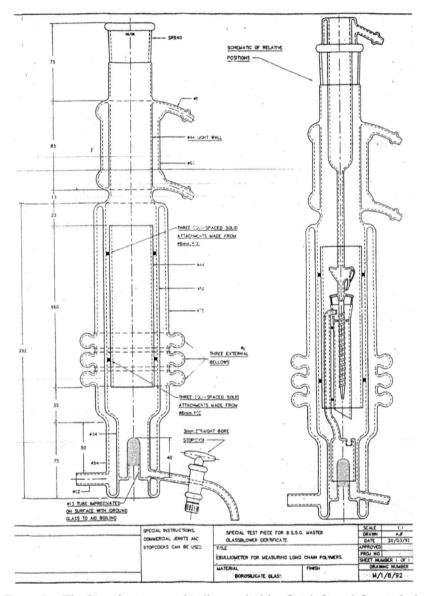

Figure 2.   The Line drawing was kindly supplied by Quick fit and Quarts Ltd., Stone, Staffordshire, at the request of Mr Fred Porter, University of Bristol. The then Chairman of the B.S.S.G.

fierce hot flame, to ensure complete contact between the two and melted more solid rod at each joint to be sure of a robust seal. I also made sure that I flame annealed each set of seals and balanced the temperature as I worked. The glassware was placed into a preheated annealing oven, which was well on its way to 560° centigrade.

At the base of the unit there is a small inverted dome into which was placed a heater, the surface of the dome was granular so as to give a good even heat transference, fortunately I had a supply of granules or glass frits of various grades as I used to make odd shaped sinters (filters). I chose to use porosity 2 grit. I took a metal plate some 150 mm by 150 mm which was 15 mm thick and supported it on bricks next to my burner, I then heated the plate with a hand torch, the frit was then spread on top of the plate and after doming a piece of tubing, heated the tube evenly and rolled it across the now warmed frit which was picked up on the glass, I then tapped the glass on the bench to ensure that the frit was secure, then repeated the process once more, which produced the required granular surface, the coated dome joined the shroud and tube in the annealing oven.

Thankfully, the inner body and shroud survived the annealing process, the heater dome was supported and sealed within the base section which in turn was joined to the inner body and the drain "take off" stub attached at the point of the ring seal.

Three bellows were made in what was to be the outer jacket together with the unequal joint for the bottom leg, after annealing the two were carefully supported in position. A seal was made between the inner and outer body and a short stub to the right of the seal to enable the eventual close back-to-back seals to be made.

The condenser and the top joint were made without incident, although I made a Dewar seal at the bottom in anticipation of the close joint to the main body.

The two items once annealed were "set" within the lathe and all blowing tubes attached. The careful re-heating began, I looked for the first signs of the orange sodium flame, which to me has always been the sign that the glass was above the annealing temperature, the reason being that my first six months training was all conducted in soft "Soda Glass"! I can assure you that having the experience of working with "Soda", one's warm up techniques are honed to perfection.

Once the pre-warmed glass was up to temperature, the ring burner was used to make the initial joint between the stub and the Dewar. The two parts were brought together and the seal made. Once the seal was completed, the ring burner was removed and the close back-to-back seal was "run in" with the intense hot flame generated by a WISPA jet burner, suddenly and quite surprisingly with the amount of heat that I was using the surface tension of the molten glass formed the shape between the two close seals, I immediately withdrew the flame and allowed the joint to cool a little then heated the condenser Dewar end to ensure that the seal came off at 90 degrees. A pre-formed side arm, of which I had four spares on the lathe bed just in case, was quickly attached and aligned with its neighbor. Disconnecting the glassware from the Lathe and the various blowing tubes as the glass was cooling down was a fraught experience and one which I would never wish to repeat. The entire piece was placed in a hot annealing oven, as I closed the oven door, I allowed myself to start breathing again!

Once annealed the final part of the main body was the base which would require a Dewar seal, therefore I had to attach the side tube to the main body to enable me to blow the seal, which was in itself straightforward. I completed the body by making the side seal and attaching the pre-prepared tap, all of which was concluded with the final bend from the tap at the bench to ensure that everything would be aligned.

I looked closely at the original distillation spiral and could see that it was a solid spiral wound around a hollow tube down which a thermocouple was to be inserted. The spiral was designed so that the returning condensate went through the rising vapors. I wrapped a spiral directly onto the hollow down tube at the bench using quite a hot flame and was relieved to find that a solid seal was made between the two.

Regarding the central down tube, although there is an offset vapor and return tube, the most important part is the heater

Figure 3.   Ebulliometer.

shroud which must be aligned centrally, not only between the top
cone/socket but must also fit centrally over the heater dome itself.

My copy of the 1930 Ebulliometer as designed by "The
Hollanders". My ebulliometer Figure 3.

The original when found was strewn with a vast number of ther-
mocouples with which to measure the unknown polymers' melting
point. Everything below the bellows had degraded to dust.

Look at the line drawing "as made", then the photo and see
that a portion of the center down tube is different; therein lies a tail
(Figure 4).

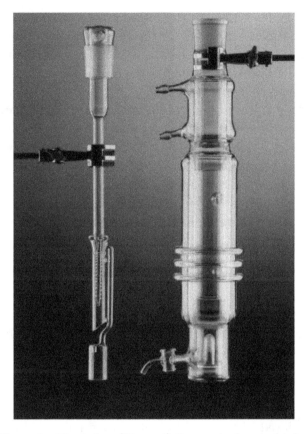

Figure 4.   Final piece (Figure 4) the repaired central tube.

Some years after making the ebulliometer, I was "volunteered" to do a four-day glassblowing demonstration for school children at the Catalyst Museum, Widnes, Merseyside, a great deal of equipment was needed such as gas cylinders, control valves, pipe work, burners, tools and glass.

The event was deemed a success as I did not cut or burn any of the little darlings.

Once I had finished, it took time to dismantle and pack away all the equipment. A chemist (I will call him David), who had been giving a Chemistry demonstration, could see that I was heavily involved in dismantling the equipment and offered to take the ebulliometer and bring it to the workshop the following Monday, an offer I appreciated very much at the time. Monday came and went, nothing! The following day David's friend and colleague rang to say that David had broken the ebulliometer and that he was too scared to bring it back!

I said "tell David to bring the glassware back immediately". Very soon David appeared and explained that he had taken the glassware home to show his children and that the glassware had slipped through his fingers and had hit the kitchen table. He kept apologizing again and again. I could see that he was upset, I put on my caring face and said "It would be best if you left now", there was a pregnant pause then David left the workshop still quite distraught, when I was sure that he had returned to his laboratory, I took the glassware and replaced the centre tube with something a little more substantial as the glassware was now for display purposes only. A little later David's friend and colleague rang to say that David was still terribly upset.

I told David's dear friend that I had already made the repair, there was a slight pause then he said, "I will leave him a little longer before I tell him" (friends can be so cruel). I said, "tell David to bring some Cadbury's chocolate digestive biscuits and that I did not want home brand and the like, they had to be Cadbury".

After lunch David returned clutching the correct type of biscuit, while we enjoyed the tea and biscuits, I eventually told David that I had completed the repair within ten minutes of him leaving and that it was in the annealing oven. To my surprise he called me a naughty name!

I mentioned at the beginning of this chapter the early mass spectrometer M.S.2. I am now aware that I.C.I. Ltd was involved with the early spectrometers as they had their glassblowing departments making various sections of the spectrometers. My head of department at Blackley, Manchester, Mr Cyril Blackburn said that when he was a young man he was asked to take some spectrometer glassware to Runcorn where it would be added to the next section.

Figure 5 showing rectangular tubing which had been redrawn vertically through a furnace suspended from the ceiling via a series of pulleys. The rectangular tubing was to fit between the two large

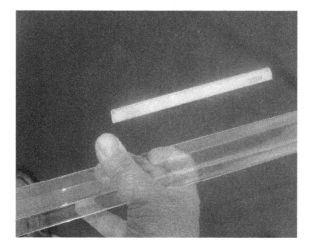

Figure 5.

magnets within the spectrometer. The redrawn rectangular tubing was 128 cm in length, 48 mm wide by 18 mm with a 1.5-mm wall thickness.

Once at Runcorn the head glassblower thought that Cyril was just a delivery boy and gave him the grand tour of the workshop after which Cyril thanked him and said that it had been a most interesting tour of the workshop as he was in fact a trainee glassblower. The mood changed immediately, Cyril was accused of trying to steal glassblowing secrets, he was then escorted off site by two large security officers.

In 2000, I.C.I. Ltd ceased trading! Fortunately for me there was a management "buyout" by people who had engineering backgrounds. They retained the technical service groups; the new owners understood that it would be an advantage to offer quality technical support and fully serviced laboratories. Much like Francis A. Freeth in the 1920s.

From then on, I had to run the workshop as a business, which was an exciting and interesting challenge. I remember on one occasion going to a company that specialized in soil analysis, I had a meeting with their Chief Accountant who had overall charge of their buying policy, he sat behind a large impressive desk within a large imposing

Figure 6.   Cup, saucer and two Cadbury's chocolate digestives.

office. After we had established what he wanted and what I could supply including probable costs and delivery dates, he then asked if I would like a cup of tea, I replied yes please but I would prefer it in a china cup and saucer. There was a slight pause, he then pressed a buzzer and his secretary appeared "please arrange for two teas Miss Brown, please note that Mr Le Pinnet would like his tea in a china cup and saucer". The accountant then turned to me and said, "would you like a biscuit"? I replied, "if you happen to have Cadbury's Chocolate digestive, yes please".

Miss Brown was so professional her face did not flicker one bit. As we waited for the tea the accountant looked across at me and put his head to one side and said "explain"?

I told him of my experience so many years ago at Manchester University and my work on the mass spectrometer, he broke into a smile and started to laugh as he brought his chair around to my side of the table. From that moment on, we had a good working relationship that lasted for many years (Figure 6).

Chapter 18

Working to the Limit

Ian Pearson

Introduction

What is this limit I am referring to? It cannot be measured with
rulers, gauges or callipers, but is the most vital dimension in any
creative process, including scientific glassblowing. It is our personal
feelings which encompass confidence, courage, failure, rejection, hate,
love, anger and patience. We all need these characteristics, in some
way more than the ability to "just" join two pieces of glass together.
We may have learned how to construct a Liebig Condenser in
a profitable timescale and to match the exacting measurements
according to a drawing, but we are not machines so how do we
interact our human emotions with the production of a physical
inanimate object? In this chapter, I will explain my approach to
the subject using three specific examples of work, two of which are
scientific glassware, and one is an item of artistic glassware, part of
an installation.

My Background and Experience

It is true and only natural that our actions are governed and
developed by our experiences. So, it is with me. I have broken many a
glass item against a wall during my apprenticeship, which was over 50
years ago. Thankfully, my destructive days are over, although I am
still having a smashing time with glass! I have learned to control
my relationship with glass so that the positiveness of working with

such a perceived fragile material remains to the fore. Unfortunately, I learned the hard way through my own experiences without anyone asking about my well-being before I started. Years ago, no-one asked how one was feeling before commencing work. Thankfully, this has changed, and my current experience has highlighted the importance of mental well-being.

In teaching basic lampworking skills at North Lands Creative in Lybster Caithness,[1] I have had students who needed reassurance and comforting, which became a separate skill from the expected requirement of an instructor of just knowing how to do the job. Quite what exactly is the job has expanded these days and it's not enough just to show a person how to say make a test tube but to enquire whether they are fit and happy to carry out such a task. Is making an item within their limits and is it possible to extend their limits? Part of this is to accept that failure is good and can be the best way to learn. This should not be taken to such an extreme where a physiologist is on hand to provide a shoulder for a trainee to cry on and allow the tears to extinguish the flame!

Example One

It's just a Mercury Diffusion Pump, said my supervisor. Aye, right I thought, sure it is. Of course, it wasn't. It was a test to see if I was experienced enough to be a grade 1 bench worker. I was employed at Jencons Scientific in Hemel Hempstead at the time and keen to earn more money and naturally be a better glassblower. The two don't necessarily follow each other! Just one hurdle to jump over and proving I could make a diffusion pump was the test and hence pressure on me. How would I cope did not enter the equation. It was all about whether I had the manipulative dexterity to carry out the various tasks of internal and external seals combined with the patience of grinding special and precise angles on glass. It is just a series of small jobs all joined up and if you get all these little bits correct, then the overall effect will be fine. It seemed so easy, but if so, why was I so nervous? Age and inexperience played a large part.

---

[1] https://northlandscreative.co.uk/.

I was in my twenties and only been with the Company a short while. Living in a new town not of my birth, I felt somewhat alien to my new environment. Thankfully, I was made to feel at home and at ease. First lesson then is do not start a job if you do not feel comfortable with yourself and your abilities. It is perfectly OK to turn work away (ideally by referring to another) rather than take on something which one cannot complete. That is a very fine line decision though and I have taken on a few jobs that have tested me, but I have learned more from trying and failing rather than not accepting the challenge.

Looking at the technical aspect of constructing the pump, what surprised me was being taught to use a small flame to fuse the top seal which has a large diameter. Initially, I thought of using a large flame and heating a larger area of glass with lots of blowing and stretching. It was such a pleasant surprise to realise I could create a superior finish by using a small flame to heat the minimum amount of glass tube to avoid distortion. I have used this method from time to time and offer this as an example of how to welcome alternative ways of working, which initially may seem strange.

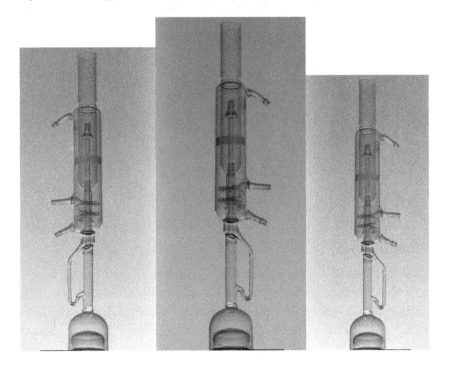

That was over 40 years ago, but I still practice the mantra of ensuring I am comfortable before I start most jobs. I have kept that diffusion pump pictured to remind me that I do have the required skills. Always good to focus on something physical to channel our positive thought. I suppose that is why there are so many icons in churches! The pump looks surprisingly good for being over 40 years old and even the scratches and dirt do not detract from its significance in my career.

The following is an image of Ian Pearson and his diffusion pump "apprentice piece" of which he is rightly proud.

Example Two

I am not a great fan of lathe work, mainly due to all the setting up time that I regard as wasteful. However, when working at Dounreay for the organisation then known as the United Kingdom Atomic Energy Authority (UKAEA), I could not escape this type of work.

If a customer required a certain piece of apparatus that required an element of lathe work, how could I refuse? I actually tried doing as much work as possible on the bench using all sorts of contraptions to avoid setting up glass in chucks. This is either amazingly stupid or a sign of fear, especially when I had some of the best lathe equipment available to me. I reasoned that it was because I did not feel in total control of the glass and it was all too mechanical for me. I felt like a machinist, which I wrongly judged as unskilled. So I had to educate myself since I was working on my own and had no one to turn to. The multi-socketed lid pictured is typical of some of the lathe work I did at Dounreay between 1981 and 1993. I found that I would spend an hour setting up all tools and lathe with the glass positioned appropriately only to run out of time. Key obviously then was to plan when I felt at my best to do such work and it transpired that around 10.30 am in the morning saw me at my peak. The main ingredient I experience which caused me anxiety was indeed time, but it was more linked to how long each join would take and how to keep the whole object warm so as to avoid cracking. Of course, I forgot about the problem of the glass falling out of the chuck half way through the job. Nowadays, much attention is given to risk assessments and I suppose in my mind I was carrying out a risk assessment every time I started to plan some lathe work. There are so many things to keep an eye on when you switch on the lathe that one distraction followed another. This got to the point that when mistakes happened, I blamed everything else other than myself! So I had to give myself a good talking to, focus and be prepared for every eventuality as well as allowing enough time and making sure I started the job when I was at my peak. Not forgetting to make sure the radio was on and that I had been to the toilet beforehand. One positive experience that I really enjoyed when doing lathe work was stepping back from the lathe as the glass was warming up with large busy flames or being flame annealed. I realised I had to increase this enjoyment factor in myself if ever I was going to address the issue of my fear. Knowing that I could be a successful lathe worker and that it was indeed a highly skilled job put me well and truly on the right road.

Example Three

One technique that I had been keen to experiment with and master was the fusing of rod onto tube to create textured patterns. I needed a reason to try this so welcomed a commission of ballet shoes which involved "running" rod through the flame onto the hollow shoe to make the laces effect. Key to the effectiveness of this was a steady right hand. I had to make sure I was in the right frame of mind to be relaxed enough so my hand would not shake. I also had to be determined and precise with my hand movements and achieved this by being focused. Sometimes experiences outside the glass workshop can play havoc with productivity. Such was the case when I attempted by first run in with the ballet shoes. An argument in a nearby supermarket was still playing havoc in my mind and thus affected my concentration with potential consequences of not only destroying the ballet shoes but burning my hand. Again, I blamed everyone except myself when it could only be myself that was not in control of my actions. I learned the hard way, but at times experiencing the hard way not only is the best way, it is the only way!

I completed the commission, and it was so well received it appeared on the front cover of the Scottish Glass Society's Journal for 2018. More importantly than that, I used the technique again for another commission of Cowboy boots. I detect a trend here and have successfully attempted similar work without putting my foot in it!

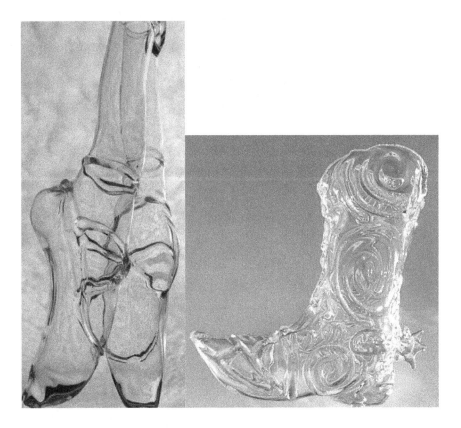

The strange thing is that after completing my sculpture of the ballet shoes I looked back over photographs of my commissions and found a picture of a glass figure I made about 20 years previously using the same technique. So, because I had forgotten that I had used the technique before, I approached the subject with a fresh pair of eyes and hands. Memories are not always helpful and can unfortunately get in the way of creativity and enforce prejudices. Always, always it is best to keep an open mind. I have worked with artists who have given me work I thought was impossible, but it did not stop me trying and it worked with excellent results. The most extreme example of this is a sculpture I had made fusing borosilicate glass to an old pair of scissors that were covered in rust. The glass stuck to the metal so well that it refused to be removed and I sold the sculpture!!

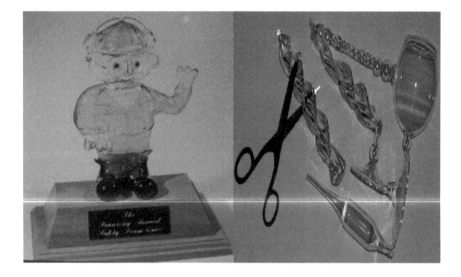

There are other limits one must consider when working with glass using a bench burner and these relate to size, chiefly the maximum diameter that can be blown. Of course, this is nothing to do with personal limits although without recognising our skills, can we tackle such seemingly impossible jobs? Such was my initial reaction when asked to make a decanter which had to contain a 750 ml bottle of wine. The shape demanded a diameter over 100 mm and all I had to work with was a "Rotajet" fuelled by Propane and Oxygen. I thought I would give it a go, but acknowledged that I would not be able to blow up the glass and keep a robust wall thickness using such a small flame. I approached the job, pretending it was an experiment and it did not matter. All mind games of course, but sometimes these so-called games should be taken seriously. After a few hours of heating, blowing, cooling and measuring, I created the final piece,

which exceeded my expectations and pleased the customer to the point where he gave me a tip!

Conclusion

All conclusions are based on my experiences and they will be different for everyone as we are all individuals and experience the same things in a unique way. Nevertheless, this doesn't stop me listing the following bullet points which have helped me as they may offer comfort to others.

- Know and appreciate your abilities and skills. These may not be limited to glass working. Some of the best glassblowers I know suffer from insecurity and lack of confidence. The opposite can be true, and I have met some very average glass artists who have disgusting amounts of confidence. Whatever skills you have, acknowledge them and use them in your work.
- Recognise that rules and guidelines can be broken. Traditional methods have their place, but if no one made changes, then we would all still be trying to bend glass rod over a candle flame!
- Push yourself but also take a break. Life is all about managing the highs and lows. I have taken on jobs I thought impossible but

ended up creating a masterpiece according to my customers! I relax by spending a few hours just messing around on the flame. There are those that treat glassblowing as just a job and while there is nothing wrong with that, it is not for me. I am lucky and thankful that I have such a skill but also recognise that it is me and only me that can care for it and develop it.

- Separate what you do into two groups. Those jobs that you can do with your "eyes closed" and those that really do test every muscle of your experience. Do not waste time concerning yourself about the little things, otherwise you will have no time to attend to that which matters. Not everything in life must be a drama! Note, this is not a view that everyone shares.

- Accept mistakes and failures as lessons to further your career and fulfil your ambitions. It does not matter when working on a piece of glassware in the flame that it cracks. What is important is that you have the courage to seek where this crack is and can repair it. After over 50 years of working with glass, I know I am only as good as my last repair.

- Do not hope or pray things will go right as they ignore your capability to plan a job through from beginning to end. Hope is a sign of desperation. I never hope my work goes well as I have expectations that it will as I have carried out assessments that it will according to my plan. That is confidence for you.

- Allow time and plenty of it. If you think a job will take half an hour to do, then allow three quarters of an hour. I have experience of working to deadlines and with the piece work method where you were paid for what one made. This style of working does tend to encourage short cuts and produce items which could have been of better quality. However, the techniques learned were applied later in my career to great effect. Sometimes joining a sidearm in one action is better than taking longer, especially better for improving one's mindfulness!

- Create the most comfortable environment for yourself as an individual. Some like music while they work, others demand silence. You are in control so what suits you is the right to do. I prefer the radio on as I like to listen to another human's voice talking about everything except glass when I am working with glass!

- Be true to yourself. I can work and create with coloured glass but choose to work with clear glass. This is my choice, but unfortunately there are critics who judge me as a lesser person because of my perceived failure to embrace the wide range of the colour spectrum.

- Understand the material you work with. Remember it's glass and not metal or wood, so treat it accordingly. This is particularly true with annealing. There is a view that covering a glass item with soot helps the annealing, but, in my view, this is a cover up! I have used strain viewers all my life is making sure glassware is stress-free by examining glass before and after work has been place in an annealing oven. Flame annealing does not have to involve soot. For the first three years of my training, the company I worked for did not own any ovens and we annealed everything in the flame. Not a single issue of concern arose using this method.

- Make friends with your mind and when in the middle of a complex job, start chatting with yourself and pretend you are a surgeon saving someone's life by carrying out major surgery. This approach really focuses the mind.

- Do not be afraid to applaud yourself when completing a piece of work that you are proud of. Many a time I have spent ages on a piece and finished it by placing it in the oven, then running around the workshop clapping my hands. I work on my own so no one can see me. What happens in Ian's workshop, stays in Ian's workshop!

Postscript

During the time that I compiled these thoughts, I was actually challenged to my limits when my workshop wooden floor gave way at the same time as the ceiling fell in and, to add even more pressure on me, the electricity supply ceased. I was ready and close to giving up when I remembered my own advice and thought I could not write guidance for others yet ignore it for my own circumstances. This was in 2020, which will be remembered for the most challenging times ever across the world, especially for those in business, but regardless of status we each face difficulties and knowing how to deal with stress is paramount to our well-being. Working to the limit can seem like giving up at times, but never give in!

Chapter 19

A Brief Guide to Neon: Cold Cathode Making*

Julia Bickerstaff

This chapter presents the equipment, techniques, thought processes, various gasses, glass to metal seals and filling procedures related to cold cathode making.

We are looking at the manufacture or making of neon tubes in this text, not the technical workings of a neon tube, but to understand the practical methods you will need a basic knowledge of how they work for best practice.

"Neon": The use of gas-filled tubes containing either 100% neon or a proportion of neon gas and others such as Argon and Mercury vapor. When a sufficient electrical potential difference is placed from one end of the tube to the other, via electrodes at either end, with a very small current of electricity, an arc is formed which ionizes the gas. This changes the gas state into plasma and thus emits light. A typical current might be 20 mA but not limited to that current; it can range from 8 mA for a glow up to and over 200 mA for a very bright light. The inert properties of the gas or gases used means that the majority of the energy used produces light and very little heat

---

*All images and text are copyright and the property of Julia Bickerstaff except image 5 that has been credited to Kansas Historical Society. No permission is given to copy or reproduce without the copyright owner's written consent and approval.

and noise often associated with the plasma state of matter. The neon tube is technically known as a "cold cathode lamp or tube".

Anatomy of a neon tube: They are relatively simple devises made up of a sealed glass tube with a metal electrode at either end with protruding wires for connection to an electrical source. The tube is self explanatory. The name "electrode" is given to the assembly at either end. The "electrode" is made up of a glass envelope (a short tube of glass) which contains the actual electrode inside. This is usually made from iron and is formed by "forge cup extrusion" often called the "shell", though an electrode can be just as effective by a simple sheet rolled into a cylindrical form. The idea is to obtain a specific surface area to emit electrons from, which is directly proportional to and measured by the current used to light the tube. The shell is welded to "lead in wires" or "lead wires", which are usually a nickel alloy, coated in copper called "Dumet" wire. This will bond and have the same expansion coefficient as the glass, which is "pinch" sealed to it, meaning the vacuum inside the tube will be maintained throughout the lamp's life. There are then "tail wires" which are attached to the electrical supply at either end of the tube via a constant current transformer. There are two electrodes on a neon tube, one at either end of the same size. The glass envelope of the electrode is butt joined (we will cover that later) in one way or another to the glass tube after it is bent to the desired shape (Figure 1).

I think the historical roots of neon making can be split into two trains of thought. Neither one is better or worse than the other. Both methods were born out of either tradition or necessity.

We can name these two methods by the continent they were practiced in originally: The European method and the American method. We will look at European methodology first.

Europe was the birthplace of the world of neon lights. The gases neon and argon and others were discovered at University College London by William Ramsey, Morris Travers and Norman Collie. While the first 100% neon "filled" glass tube was made by Norman Collie, little consideration was given to the use of such a tube and it was laid to rest as a discovery. It was a French entrepreneur

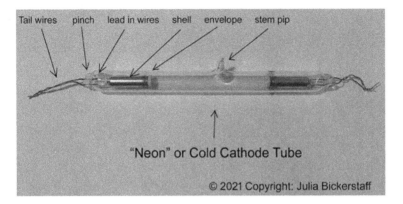

Figure 1.   Glass envelope and pinch seal.

and scientist, whose name is synonymous the world over as the "inventor" of the neon tube: Georges Claude, who capitalized and developed the gas tube to a level to which it was commercially viable as a lighting method, better than any other on the market at the time. There is a coating applied to the inside of the electrode shell to improve its efficiency and processing, this coating was Georges Claude's invention. What we need to remember is that the only people with the skills to work glass tubes were scientific glass blowers (though not necessarily freehand bending of glass tubes). The making of these neon tubes fell to them. The result, of course, was that they trained workers specifically in the making and then on to bending of the glass tubes; this was in the demand period prior to the outbreak of the First World War. The skills acquired by these new glass workers developed and excelled with practice, so they could manipulate or bend glass tubes relatively easily and quickly. They would use the detailed jointing methods and craftsmanship employed by scientific glassblowers to fit electrodes. Accuracy and detail is taken to extreme with exact fit and appearance maintaining wall thickness and durability.

Typically, the glass used was Soda Lime and Borosilicate.

Most neon glass workers today use soda lime glass with additives to make it more tolerable to work (via a larger/longer transition temperature range). This form of glass is branded and known as

"Euro-light" or "Lead-free", simply because it behaves in a similar way to lead glass. To work this form of soda glass the burners use a combustible gas like methane (natural gas), butane or propane and forced or compressed air at around 0.7 Bar or 10 psi to enrich the flame. The reason for using soda glass is because it is comparably a lot cheaper than borosilicate glass and it melts/softens at a much lower temperature of around 800°C and the flame is quickly adjusted for work rate with only two controls. Some neon workers today in Europe still work with borosilicate glass (a brand example is Pyrex). To be able to reach the temperature needed to melt borosilicate, you will need to enrich the flame with oxygen to achieve the required temperature. Using oxygen is familiar to scientific glassblowers as a lot of scientific glass work is in borosilicate. Here, I am concentrating on Neon-making specifically in soda glass. There are generally three types of burners used for the European glass work methodology. These are the burners I use in my workshop.

These three burners are:

Figure 2.   The bench torch.

The bench torch started out mounted on a bench as indicated by its name as you would see in any scientific glass workshop (see Figure 2). However, if you stand at the torch rather than sitting to work, you can use the whole of your upper torso to handle the glass. Added to this, if the bench torch is mounted onto a stand remotely from your bench, you have a much greater area around the torch to wield the glass tube length.

The hand lamp is used for the more traditional glass work of jointing tubes both as butt joints, 90degree joints and T-joints, etc. for electroding (attaching the electrodes) and extending/joining tubes and fitting capillary glass tubes to facilitate the processing of the glass tube (so it lights up). The hand lamp is almost always used sitting down at the bench (see Figure 3).

The notable feature of the burners using the "European method" is that the bench torch has a single flame, as does the hand torch.

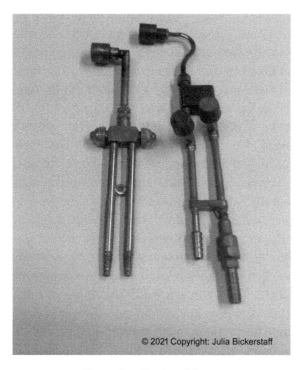

© 2021 Copyright: Julia Bickerstaff

Figure 3.   The hand lamp.

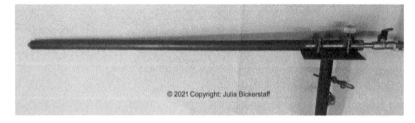

© 2021 Copyright: Julia Bickerstaff

Figure 4.   The ribbon burner.

The combining feature of the European method and American method is that the ribbon burners are identical (see Figure 4). The principle is that the flame is a long stretch flame approximately 15 mm wide and adjustable in length. There are a number of different sizes (adjustable lengths) available from as short as 150 mm up to 900 mm and longer. The size I use is an 800 mm burner used for slow bends/large radius bends (which I will explain later).

Now let us turn our attention to America. The sale of the first neon tube sign by Georges Claude's company received a marketing media triumphant embrace, although it did so notoriously! After all, nothing like this had ever been seen before. The requirement for such an advertising advantage was thrown into high demand. The obvious step by Claude's company was to train people specifically in making neon. The demand had been such that new techniques evolved and the American method came about. No sitting down to work, all work is carried out standing, including attaching electrodes, which incidentally is done quicker by butt joining the electrode to the tube and then bending the glass over in a turn back or angle bend (I will explain later about bends), thus speeding the process up and making it simpler. Details were not so necessary since most American neon was installed at high level and detail was irrelevant. Speed was the driving force. They employed different types of burners, too.

Again, three types of burners are commonplace:

The flame used was not just a single flame, why do that when multiple flames pointing inward would provide fast heat all-round the glass tipping torch/hand lamp? (Figure 6).

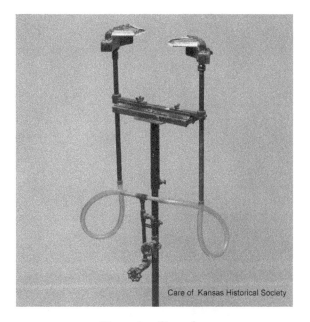

Figure 5. Cross fire.

Figure 6. The hand torch.

Unlike the European single flame hand torch, the American method implements two flames on a "U"-shaped tube pointing at each other in the centre of the "U" shape.

The ribbon burner is essentially the same as in Europe, a good design. Some though are made from a cast block (Figure 7).

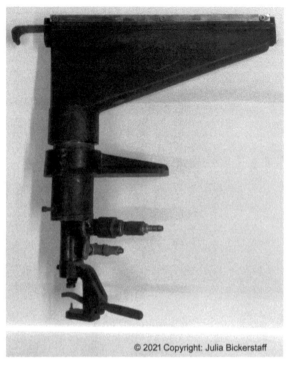

Figure 7.   The ribbon burner.

There is one other aspect to consider, the method for blowing the glass. Traditionally in Europe the neon bender/maker would blow directly into the open tube to change internal pressure of the tube, thus inflating it. Remember the other end of the tube must be bunged, while bending it or straight after bending it. The only time a blow pipe and swivel (Figure 8) are employed is for butt joining or fitting electrodes. However, in the interest of speed and economy the American method employs the blow pipe and swivel for all manipulation blowing bends, electroding and butt joints.

Today we find globally a merged and evolved use of both techniques. The burners used do still have the unusual trend of single flame and multiple flame heads. Perhaps this can be put down to tradition. Come to think of it, some neon makers only use a single flame hand lamp to work the glass, horses for courses.

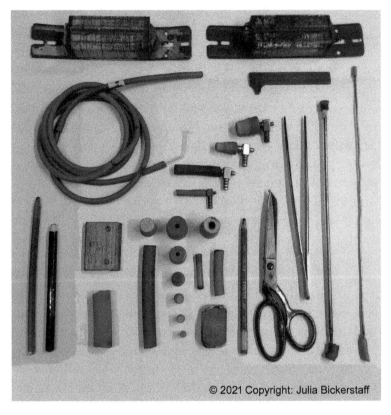

Figure 8. Swivel and various tools required.

The other tools you will find useful and in the possession of all neon benders/makers are a stout pair of tweezers, a glass knife/carbide blade or fine engineering file, a very soft pencil or china graph for marking the glass, lots of bungs and an array of home-made tools for reaming and handling hot glass. Also, tools for taking powder out from inside the tube. You cannot butt join or fuse electrodes onto glass with powder coating at the joint. This is because it will crack after cooling (Figure 8).

Now you have your burner/s, it is time to select your glass. Typically, the diameter is dictated by the detail of the work. One of the most important things to consider is lamp life. Selecting the largest diameter glass, you can manage and work for the piece to extend the lamp life. There are formulas to calculate expected lamp life. The rule to remember is that the larger the diameter and the longer a lamp, the more volume there is inside, which is directly proportional to lamp life. So, the tiny neon signs made in 6 mm diameter glass will not last as long as a large neon installation. Therefore, neon tubes are so good for lighting large areas because you can use large diameter glass, for example, 18 mm, 20 mm or 25 mm, and achieve a really long lamp life. There are many diameters in between. Typically, 8 mm, 10 mm, 12 mm and 15 mm. Colors other than the red of pure neon, which is recognizable by its rich crimson red in clear glass, are achieved with powder coatings or phosphor powder. This changes the wavelength of light from Ultraviolet C within the tube to a visible color. For example, pink, blue, whites, green, etc. The whole spectrum is available and a vast array of full spectrum whites, too.

It must be said the majority of techniques for glass manipulation are achieved through practice; it is something which is learned as opposed to been taught. Imagine a musical instrument: You can make noises quite quickly, as you can manipulate a shape in the soda glass tube after a short space of time; within minutes. But it takes several months of hard practice to make a tune; to be able to bend glass to an effective level. But it takes up to a decade to master the instrument and the sound to be of an orchestra quality; to be able to have full control of the glass and be the master of the skill set required. Practice makes perfect, think in terms of months and years not days and weeks.

The best advice I can offer is to take your time and consider making one bend at a time in a single piece of glass and then cutting and joining the various parts you have bent together to create a letter, then if you are creating a word, join the letters you have bent to make up your word. There is what I call the "magic letter"; the Capital letter "R". The reason for this is that the capital letter R contains four bend types (Figure 9).

The *angle bend,* the *slow* (large radius bend), the *turn back* or return bend and the *drop bend,* which have the 3D element of bringing glass from the back to the front, synonymous with neon

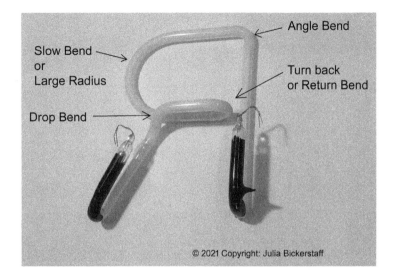

Figure 9.   Four types of bend.

tubes. Eventually and quite quickly you will find yourself bending complete letter forms. The other advice is start practicing with smaller diameters, but not too small, try 10 mm first. Because the wall thickness is around 1.5 mm in all tube diameters, the tube will inherently support itself better with the smaller overall tube diameter. When you are "electroding" your tube, this is the act of fusing/butt joining your electrodes to your tube. You will need to add a tube called a "stem" to the side or end of your electrode. The purpose of this is to provide a means to suck/evacuate all the air from inside the tube, clean it and then add the gas of choice for illumination. This is a little tricky, I use 5 mm diameter stem glass, but most neon glass shops use 6 mm as a stem. This needs to be sealed on to the glass of the electrode, usually just after the metal electrode shell. See Figure 1. If you are adding mercury to your tube, then you will need to add a small "inflation" or bubble called a mercury trap. Mercury is added to the tube after you have processed it, see what follows. I would recommend you take professional advice and precautions before considering using mercury. You need to meet with your current local regulations and controls, etc. The size of

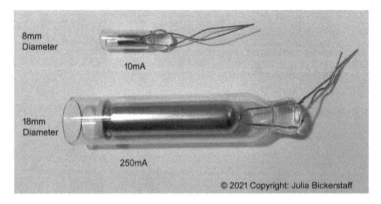

Figure 10.   Electrode sizes.

your electrode is determined by the brightness/running current in milliamps of the tube. You can run bigger electrodes at lower current, but not the other way round. The larger your electrode, the longer your tube will last (Figure 10).

Considering you have now bent some glass tube and been successful to join an electrode on either end with a stem attached and it is completely leak-free, it is time to process your tube to light it up.

Essentially, what you are aiming to do is clean the inside of the tube and evacuate it completely. There are two methods for processing neon tubes: Electron bombardment and Oven heating. Both methods require a high vacuum pump of one type or another to evacuate the inside of your tube. The "Bombarder" is by far the most commonly used, but many larger neon glass shops also use oven pumps. Historically, an oven pump, as the name suggests, is a large chest in which you place your tube or tubes and attach the stem to a manifold and evacuate the air at the same time as heating the oven. The electrodes also need heating, the oven is not sufficient to do this, so use an induction coil to heat the electrode shells till they are glowing cherry red. This will ensure they are clean and the coating inside is "activated". The neon tube will last a lot longer the cleaner it is. The bombarder however cleans the inside of the tube and heats the electrode shells at the same time. This is a complex apparatus, but it is efficient at processing tubes individually or as pairs. Essentially, what you do with a bombarder is to attach your tube to the manifold and begin drawing air out. Once you are down to around 6 or 7mBar, depending on tube length and diameter, you can light it with the air inside. Because air has a high resistance to electricity, the air will heat up and in turn heat the glass tube. This heat will vaporize any impurities in the tube so they can be evacuated out. The electrodes however are heated by the rising current in the tube. This is achieved by increasing the current passing through the tube using a coil transformer which has its current adjusted and controlled one way or another. My bombarder uses a chocking coil, as you withdraw the iron core from within the coil it allows current to flow. This can be anything from around 100 mA up to 1000 mA (1 Amp), does not sound much does it? But when you consider the voltage of the bombarder transformer is between 10 kV and 20 kV, it does amount to enough to heat up the electrodes. With a delicate eye and control of the apparatus, bearing in mind the diameter of the tube, the size of the electrodes and the length of the tube, it is a balancing act of control to achieve the perfect result of cleanliness. When the tube and electrodes are hot, it is time to immediately evacuate to a complete vacuum. This removes the impurities. If you are satisfied the tube is clean and suitable for the filling gas of choice, for example, 100% neon, then you can go ahead and fill it, but only after the tube has cooled down to room temperature. Filling a hot tube will only serve ultimately to give you a tube with a too lower pressure when it has cooled. The fill pressure is denoted by the diameter of the glass and the type of gas you are putting in. Manufacturers of raw materials provide charts to help with this. It can be anything from as little as 7 millibars for 25 mm diameter tubes up to 30 millibars for small diameter tubes. Something to bear in mind is that the more gas you use, the longer the life expectancy, though this is a trade off of light output and needing a higher voltage to light the tube. It is also worth putting a small percentage extra gas inside a very short tube, since lamp life is directly proportional to its internal volume, especially if a small tube is placed in the same circuit as longer tubes. Should you lack the confidence that the tube is clean, then you can add an extra cleaning procedure using a small amount of Helium. Helium gas has a lower resistance to electricity than air but higher than argon or neon so will heat up gently. It is ideal for checking the cleanliness of tubes as it lights a consistent creamy white colour. If it is not a consistent creamy white colour, then the tube is dirty still. Leaving it lit with helium inside can clear the impurity and complete the process. Or alternatively, after the tube has cooled let a small amount of air in and start the process again. The electrodes will not survive multiple bombardments though, so be warned. Clean first time is always best. Sealing the tube off involves heating the stem at the point of entry to the tube leaving a stem pip (Figure 1).

Let us look at installing your finished tube or tubes. Modern switch mode power supplies at high frequency were a great development for installing neon. They automatically alter the voltage to suit your tube's needs and they have built-in protection to switch off the supply should damage occur or a fault. On the other hand, traditional transformers, known for their heavy weight and size, must be calibrated correctly for the tubes you are lighting. Please do not be tempted to use short cuts and use a high voltage 10 kV transformer where only a 3 kV transformer is needed. The installation will suffer, primarily the transformer. It is similar to a short circuit state and will generate unwanted and very unnecessary heat.

I hope I have given you a positive and informative although a very brief insight into the wonderful world of neon lights. Just to note, the history goes way back before Ramsey or Claude. It is worth exploring the work of Heinrich Geißler, William Crookes and Daniel McFarlan Moore among others. I have included a photo of one of my more unusual neon tubes so you can see you are not just restricted to straight-sided tubes (Figure 11).

There is a rich pedigree in the world of glass, light and its uses.

The world of neon signs has changed a lot. From a highly desirable advertising method through into the interwar years and on into theatre land and films to the more brash world in which we now find ourselves needing to refer to the century old craft as "real" neon signs because of the technological advancement of the LED (light

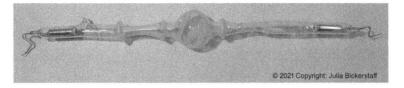

Figure 11. Unusual neon shapes.

emitting diode) encapsulated in set silicone gel which can be made to mimic neon effectively and are now prolific. Many materials have been tried to replicate the photonic glow of neon tubes. Even with the most recent advances in LED technology, silicones and plastic components still fall short, in my opinion, though they do offer a cheaper alternative.

It's a strange profession which nestled in the attics of sign companies for decades into the loud 1960s and is now seen to be used in artistic expression and personal statements whether pictorially or by the medium it is historically linked with, text and the written word. Though neon is expensive if made right, it is still used and embraced by the discerning eye. There is a tentative future but very different to the past neon enjoyed. The neon glass shops are closing the world over, but with a strong online digital presence this makes it seem as though it will never go away. Which I feel is a positive outlook for this, the original energy efficient lighting source.

Chapter 20

Bursting Disc Failure Detector Tube

Paul le Pinnet

Fellow of the British Society of Scientific Glassblowers

Re-Drawing of Multi-Walled Tubing

A demonstration of the following method took place at the British Society of Scientific Glassblowers Symposium in Telford nr. Shrewsbury, Shropshire, England, in September of 2000.

The technique may well be applicable to most small-scale re-draw applications, the making and use of De Khotinsky wax should be of interest as may be the described glass silvering method.

The Glass Services team based at Runcorn in Cheshire provided a universal service to virtually all the major Imperial Chemical Industries chemical plants in the north west of England. One such service was to manufacture and supply the all glass bursting disc failure detector tubes for those discs that were located in inaccessible parts of the chemical plant where failure to detect might lead to fugitive emissions of toxic substances into the atmosphere. Such releases might not only be expensive but might also exceed the environmental legislation and the Company's own environmental policy.

The detector tube must be made from an inert material and be sufficiently robust to withstand Chemical Plant conditions for a period of up to one year and yet be fragile enough to be broken by the shards from the exploding bursting disc. The completed detector tubes had a diameter of 6.8 mm and a usable length varying from between 250 mm to 560 mm. (Figure 1).

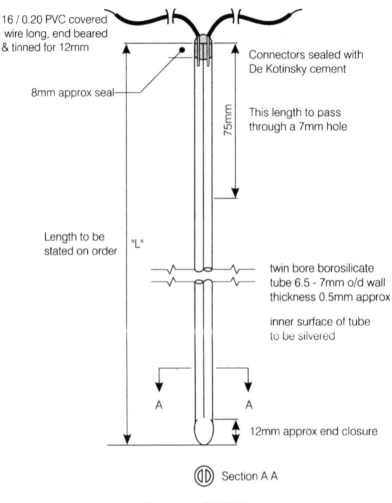

Figure 1.   BDFDT.

The materials required: 34-mm heavy-walled borosilicate tubing which has a wall thickness of 3 mm.

3-mm-thick Borosilicate flat plate which is to be cut into lengths of 100 mm by 28 mm on the diamond saw. A trial strip 28-mm-wide is cut and inserted into the 34-mm tubing to produce an equal cross-section. The edges of the plate can be left "as cut" or slightly radiuzed to give a close-sliding fit within the tube.

Figure 2.   The carbon block.

Once the size has been confirmed and noted, the plate is withdrawn, and a substantial spear point pulled and then allowed to cool.

Replace the plate within the tube while holding the spear point in the left hand and pull a second spear point to the right of the plate, at this moment in the procedure one does not want the plate to "stick" within the tube, position the left side of the flame approximately 25 mm to the right of the plate and incline the left hand downwards, thus stopping the plate sliding toward the melting glass (Figure 2).

Substantial straight spear points either side of the plate are essential as during the drawing operation a high degree of tension is produced.

The body is warmed thoroughly in a clear gas and air flame, one should be able to hear a clicking noise as the glass is rotated, this should re-assure you in that you are still in control of the situation as the plate has not inadvertently "stuck" within the body!

Once confident that everything is warmed through, concentrate a small fierce gas oxygen flame on the right-hand corner of the plate which is sealed against the inner tube. While still molten, ensure that the plate is central by watching carefully and controlling the slight movement of the plate by the angle at which you hold the glass.

Figure 3.   Slight overblow required.

Then and only then seal the adjacent corner. The tubing is then heated along its length against the edge of the plate, one can see a clearly defined contact line once the plate and tubing are sealed together.

A carbon block, which can be seen in (Figure 2), is positioned in front of the burner. Once hot, the glass is drawn over the carbon block to ensure full contact has been made. The second edge is similarly sealed within the tube. At this point a slight overblow is required (Figure 3).

It is important that both plate and tube are always kept warm to avoid surface cracks on the heavy-walled tubing and to keep the temperature balanced. Once one is satisfied that the plate is completely sealed, the whole length of the body is "soaked" in a hot gas oxygen flame until one is sure that both plate and tube are evenly heated throughout.

The glass is withdrawn from the flame and the drawing process begins with a slight pull to get things started, at this point it is important to have a beautiful assistant standing to your right, as

Figure 4. The beautiful assistant.

you start the drawing process, while ensuring that the plate remains vertical. Once one has extended the first part of the draw, the right-hand spear point is passed to the assistant. The full drawing process is complete once you both can feel the glass solidifying. It is essential that both work in unison and both keep the plate vertical to ensure that it does not twist or spiral (Figure 4).

There is so much residual heat in the glass at this point that it is important that a constant tension is maintained until the glass has cooled sufficiently such that there is no chance of sagging taking place.

Using a standard tungsten glass knife, one end is scribed and cut off! The drawn tube must pass through a go/no-go gauge. Which is a brass block with a 6.8-mm clearance hole drilled through.

"It never ceases to amaze me just how accurate the hand and eye co-ordination develops with practice".

The selected length is cut from the drawn tube using a standard glass knife, the scribe mark being made over the inner plate, then the glass tube is pulled apart in the normal manner. The tube cuts easily even though there is a plate sealed along its length.

A straight joint is made between one end of the drawn tube and a piece of either 6 or 7 mm diameter tubing, then a domed end is blown while bearing in mind that the domed end will itself have to pass through the 6.8 mm gauge. After being annealed, the inner surface of the tube is now silvered.

The Silvering Process

I will describe the method using a silvering solution as supplied by London Laboratories Ltd. It would be advisable to contact one's local supplier, although the safety aspect and general procedures will no doubt be remarkably similar.

1. Preparations to be carried out in a fume cupboard
2. Preparations to be carried out behind a polycarbonate shield
3. Safety shield must be worn
4. Rubber gloves to be worn
5. Use a mechanical stirrer (magnetic type)

Note: All references to "water" imply distilled water because ordinary tap water contains traces of chlorine — which would react with the Silver Nitrate within the silvering solution causing a precipitate of Silver Chloride, which would render the silvering solution useless.

Part 1:
1. Add 14.2 ml of ammonia (90.80 S.G.) to 17 ml of water
2. Add 12.5 grams of silver nitrate
3. Stir well until nitrate is dissolved
4. Add sufficient water to a total volume of 50 ml
5. Add 50 ml activator London laboratories MS-S-277

Part 2:

Take 75 ml of the resultant solution from the Part 1 Formula. Add sufficient water to bring the volume up to 1 l.

It is most important that no excess silvering solutions should be stored overnight.

Add sufficient water to 37.5 ml of Ma concentrate to bring the total volume to 1 l and stir well.

Add 0.5 milliliter of RNA concentrate to 1 l and stir well.

The Process

1. Clean all glassware and ensure that it is free from grease.
2. Rinse the glassware with sensitizing solution.
3. Give the glassware a final rinse with water.
4. The solution prepared as Part 2 (silvering solution) is used with the alkaline solution 1:1.
5. The Part 2 silvering solution must always contact the glass first. Although the two solutions can be pre-mixed.
6. Allow sufficient time for the solution to coat the glassware as required.
7. Pour away the residue and rinse out with water.

I have always coated the glassware two if not three times to ensure a solid dense silver coating.

The open end of the bursting disc tube is abraded with a diamond pad to ensure that the conductive nature of the silver coating remains within the double-walled tube alone. An ammeter is used to check continuity and conductivity. The resistance should be less than 1 Ohm per linear inch (25 ml) of detector (Figure 5).

The electrodes complete with flexible wire connections are first sprung into each side of the bursting disc detector tube to ensure contact with the silver coating; they are then held in place with "DE Khotinsky Cement", the ingredients of which are flaked Shellac and 5 ml of Butyl Phthalate. Dissolve the shellac in a beaker and heat sufficiently to allow it to remain molten, do not allow it to overheat, add a small amount of Butyl Phthalate and allow it to dissolve in

Figure 5.   Small gold-plated spring-loaded electrodes.

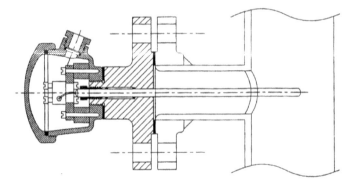

Figure 6.

the shellac. At this point, test the cement for plasticity by pouring a small amount onto a piece of flat glass and roll it while still hot into sticks about 4 ml thick. Cool it in water, then bend it until it breaks. The sticks should take 10-degree bend before this happens. If necessary, keep adding Butyl Phthalate until the required flexibility is achieved. Pour out the cement into cold water from this procedure and make the cement into sticks.

The completed detectors are once again tested for continuity and are then suspended directly over the bursting discs (Figure 6).

And an electrical contact is made, usually to a remote-control center. Should the bursting disc rupture, the glass will break and the signal will then be lost, resulting in alarms sounding. Secondary controls are then put in place to minimize damage to both individuals and control of emissions to atmosphere.

Chapter 21

Note Taking, the Resource that Never Devalues

Greg Purdy

Glass Services Ltd., Hamilton, New Zealand

I assume as you read this book you have already read Paul Le Pinnets first book *A Practical Training Method* and are somewhere on your training journey — a journey that really doesn't have an end. Rather than show you the construction of glass items which by now you are likely to have a reasonable knowledge of, I intend to share some of my experiences about recording an item's details before, during and after construction, with the view that I may make the item again in the future.

From the first day of my career as a scientific glassblower, my teacher, the late Robert Barbour (*Glassblowing for Laboratory Technicians* — Books 1 & 2), impressed on me the value of keeping notes. This advice has paid dividends time and time again and now, nearly 50 years later, that advice is as relevant today as it was then (see Figure 1).

What has changed, is how I make these notes/records. Many times, a customer has come into my glass shop with an item they want duplicated but cannot leave the item with me. My sketch pad has been superseded by my camera. Looking at a photo of the item is most often superior to a drawing as you can see how it really looks, especially joints and bends. I usually take a photo, print it, then add the dimensions in pen or pencil (much faster than on a PC)

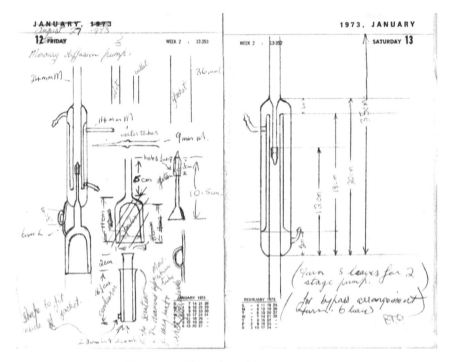

Figure 1.   The value of keeping notes.

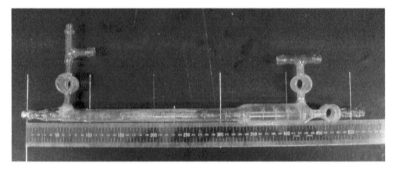

Figure 2.   The ruler comparison.

for the customer to check and confirm. It is common to hear "can you make that a bit shorter" (or longer etc. etc.). If the customer is in a hurry, I either take a photo of the glassware beside a ruler (which is dimensionally a bit hit and miss) (see Figure 2) or write the dimension on the glassware with water base felt pen then take a photo.

Ask the customer questions about the glassware they require and understand how the glassware is to be used — you may be able to improve the design or simplify its construction. CAD drawings may have fine tolerances on dimensions, e.g.: ±0.05mm automatically added by the drawing program made for items that can be cold machined, the actual tolerance, for example, may be ±5.0mm. How the item looks in the CAD drawing may not be how it is made or looks when made using standard glassblowing techniques. Experience has taught me not to hesitate to ask questions about a drawing or sample glassware — just because a customer has given a sample of glassware or a drawing of a glass item they require, does not necessarily mean this is the best way for the final glass item to look. Understand how the item is to be used and what your customer wants to achieve. Making a prototype is often the best step, especially when multiple items are required. On several occasions a customer has brought in a drawing or a sample of glassware to be duplicated, and on asking why it is made the way it is, the answer is "I don't know, it's just the way it's always been made". There is often a way to make the apparatus work more efficiently or there is an easier way to construct it, or simply just make it more robust. The notes I make on the drawing consider all changes and possible modifications.

Newly completed items of glassware can also be photographed, and dimensions added (digitally if you have the skill and time (see Figure 3) otherwise printed, and dimensions added using old school methods, then scan so you have a digital record. (see Figure 4).

Another advantage of digital photography is that glassware constructed with multiple items can be digitally assembled on your screen as in this photo of SO_2 wine testing equipment illustrated in Figure 5). The dimensions and notes were also written with a touch screen computer and smart pen, but this can be very slow and I find it frustrating.

Either way, it is up to you, but I always have a printed copy to take to the bench when I am making more of the item or to show a customer. For some of the more interesting glassware, I take

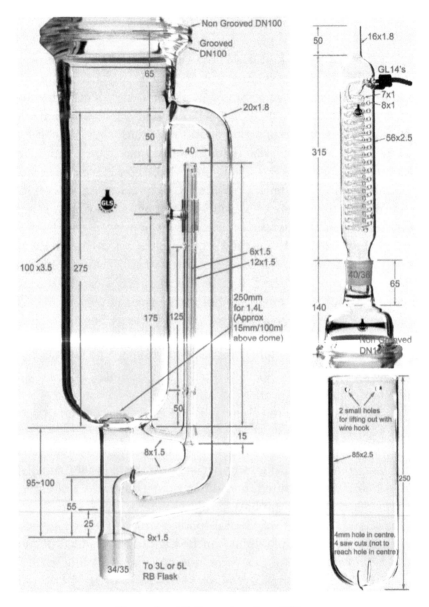

Figure 3.   All dimensions in millilitres.

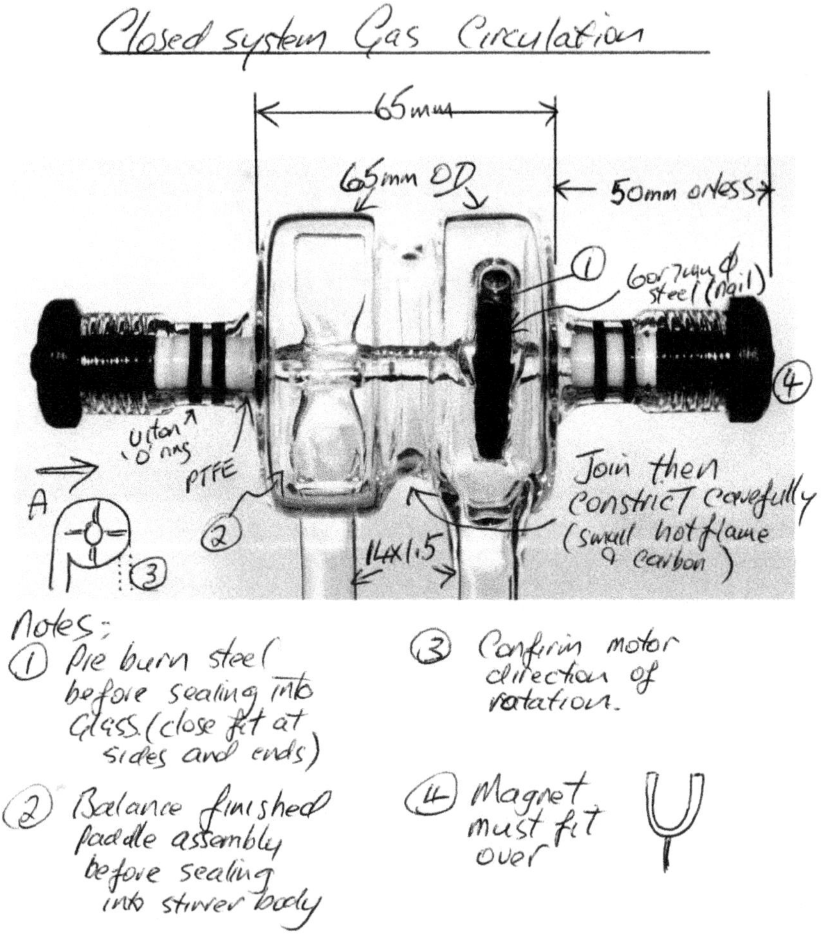

Figure 4. Old school!

two (or more) digital photos, usually only one to record dimensions, the others for larger colourful prints to decorate my workshop and customer area. Keeping notes on how you made the item is also especially important, no matter how good your memory, there's a chance you won't remember some of the trickier parts of the item's construction, even how long you cut the ground joint tube, stopcock side arm or tubing, as you know some of the tube usually ends up in

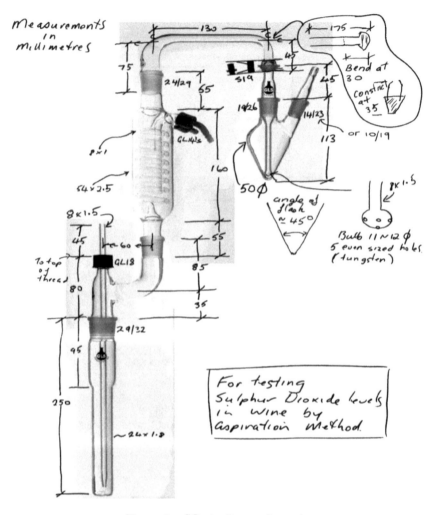

Figure 5.   SO$_2$ testing equipment.

the radius of a joint onto a bigger diameter tube or into the radius of a bend.

I sometimes take assembly and progress photos of items and the setup, especially graphite setups for profile shaping Figures 6 and 7 are the first and last of several photos of how I hold ground joints when making DN100 multi-flange tops.

Figure 6.   Multi jointed flanges prepared.

Figure 7.   Multi jointed flange tops completed.

You do not have to be an expert photographer and/or graphic artist or have high tech equipment. I have taken photos of glassware in the workshop in front of a window (facing the outside) with a colored cellophane sheet (five different colors for NZ$2.00) taped to a large wrapping tissue as a diffuser, over the window, Figure 3 is an example. A blue sky makes a good background, especially when the glassware is in the shade, layers and wall thickness show clearly. Hint — focus on an edge or side of the glassware, cameras, whether it is a DSLR or your phone, can have problems focusing on clear glassware. If you want to get more serious with your photography, consider getting an infinity light table with a semi ridged translucent bendable plexi-glass table and lighting, these tend to be better than a light tent for clear glassware as the glassware can also be lit from below or behind, once again the colored cellophane can be used (Figure 8).

In the case of glassware that is fixed in position, for example, a vacuum line, each item can be photographed before joining then another taken of the whole vacuum line when finished, a bounced or direct flash may be required.

Figure 8.   Glassware lite from below.

Freeware programs are available for viewing your photos — I mostly use a program called FastStone. It can be set up so your photos are automatically shown in screen size and is an excellent viewing program with several editing features. Another freeware program that has more editing features is IrfanView — these programs are free, but it is nice to make a donation to help with developments. I use Adobe Photoshop — not freeware — for serious editing for poster size prints. When I want to arrange photos on the same page to view or print, I use Microsoft Word and 'insert' the pictures, which can then be changed in size and moved around the page with a click and drag of the mouse once you right click on the photo and select "Wrap text option" and change it to almost anything but "In line with text." If you want more editing features, you can step up to Microsoft Publisher.

Chapter 22

The BBC, a Knight and Me

Phil Jones

Chief Glassblower, Bath University, Bath, UK

On December of 2014 the workshop phone rang and a plumy voice at the other end of the line introduced himself as Patrick Aryee, a researcher for the BBC. I wonder if you can help, he asked "we are trying to replicate an experiment from the 1950s for an upcoming science documentary we are recording for TV and we are struggling to source the appropriate glassware. This came as no surprise to me. Expecting glassware to survive 60 odd years of use in a laboratory is a tall order. However, my curiosity was aroused.

A fleeting description of what was required revealed that nothing like it would be available off the shelf, but I could make a faithful facsimile. After all, it just seemed like a couple of flasks, stopcocks, tubes and some electrodes for a final flourish. That would be wonderful if you could make this said a delighted Patrick. And a deal was struck "just one thing, we need it straight after the new year; can you make it that quickly"? Somewhat miffed that I had been distracted from my festive musings, I begrudgingly accepted the challenge.

The Origins of Life

The realization struck me that I was making something for television.... for the BBC.... something that would be viewed by millions. Further correspondence with my new accomplice revealed that they were trying to replicate the famous Urey–Miller experiment, an experiment to simulate the chemical origins of life on Earth no less! This upped the ante somewhat, I had better understand how this glassware was really used.

I discovered that in 1952 Stanley Miller and Harold Urey worked together to devise an experiment to support the theory of primordial soup that the complex chemicals required for life to develop on early Earth could be produced naturally. The theory involved the interaction of hydrocarbons, ammonia and lightning, all thought to be present in the exceedingly early days of Earth's development.

The glassware set up consisted of an enclosed system of a 2-liter flask with two tungsten electrodes attached, a condenser and an additional 500 ml flask. Water would be boiled in the smaller flask creating water vapor that permeated its way to the larger flask, where ammonia, methane and hydrogen were introduced. An electrical discharge across the electrodes simulated lightning passing through the water vapor and gasses. A condenser would cool the vapor to be recycled into the smaller flask, then to be re-heated. This experiment ran continuously for two weeks. Analysis of the resulting product revealed that up to 20 amino acids were found. And there you have it, the building blocks for proteins, DNA and hence life on Earth! It has even been discovered that analysis of vials taken from this experiment and left for years revealed even more amino acids.

A Surprise Invitation

Returning to the task in hand, the reproduction glassware was completed just beating the deadline and I still had time to enjoy the frivolities of Christmas and New Year. Job done! Except Patrick had another request "could you lend me a heating mantle, retort stand and other such paraphernalia to support and run the experiment?" I knew that the teaching laboratory technician Robert Stevens would help here, which indeed was the case. Finally, that was that I thought, Patrick was on his way, fully equipped to do his recording and I am free to resume my proper work.

Two days before filming was scheduled Patrick called, "Phil we have a bit of a problem" I immediately thought that they had broken the glassware. "We have just gone through the risk assessment and we need someone with lab experience and who can handle glass appropriately," he said. By now the teaching classes had restarted and life at the University was at full throttle once again. "I am sorry Patrick, I do not think anyone is available at such short notice," I replied. "Oh no! This is bad news. We cannot possibly let down the presenter of the program," Patrick said with his plumy voice taking on a distinctly worried tone. "Well, that is unfortunate," was my somewhat dismissive reply. "By the way who is the presenter?" I asked. "Oh, didn't I say? Sir David Attenborough. Of course, if someone could help, they would get to meet him," Patrick said, but his mind seemed distant as if trying to work out how he was going to get round his dilemma.

My Mind was now racing — Sir David Attenborough, akin to my adolescent excitement of Beatlemania, I said "I am sure that I have some annual leave I could take, maybe I could help you out, I have some laboratory experience and I know how to handle glass. I will ask my boss if its O.K." Patrick was now in a happier place.

The Gathering

Annual leave secured, everything was in place for an early start one chilly January Morning. Filming was to take place in the Biochemistry labs at Bristol University, a location deliberately chosen for its old school retro authenticity, which went with the theme of the program. I was immediately impressed by the plethora of people required for the shoot, the director, producer, researcher, cameraman, sound recordist, media assistant, electrician and so on. With military precision, everyone knew their task and set to. I was left to set up the glassware, attach the heating mantle, connect the electrodes and get the experiment running. By late morning everything was in place and the great man arrived. I asked the produce "how shall refer to him?" "Sir David will be fine". Reverence due and I was happy to oblige.

Sir David, the consummate professional, had read the script while being driven to the shoot and had made his own annotations. He took position next to the glassware. However, there was a problem with the experiment, the cameraman said that he could not get a good shot as the water was heating up a little too vigorously, it was steaming up the glassware and obscuring the view of the internal workings. "I can rectify that ," I said and reduced the setting on the heating mantle. This put me shoulder to shoulder with Sir David for a few moments. Being TV Royalty, I thought it best not to talk unless he talked first. He then asked if I was the man who made the glassware to which I answered in the affirmative. "You are a clever man, aren't you?" "Not nearly as clever as you, Sir David" was my nervy reply. It drew a polite smile and an accepting nod of the head.

We were now ready to roll. Sir David lit up as soon as "action" was called. Before I knew it, the Director called "it's a wrap". After a brief chat with the crew Sir David was on his way and after lunch, we all set to dismantling our various bits of equipment.

The crew were all very welcoming, but it is quite a different life to what I am accustomed to. I heard tales of assignments to Borneo, Patagonia and to the poles with a brush with a polar bear. The recording went on to be part of *Attenborough: 60 Years in the Wild*. Originally screened on the BBC, it is now available on DVD.

And so, my deed was done, I left with happy memories of an enlightening look behind the scenes of a TV recording, of meeting a Knight and an obligatory autograph to boot.

Which leaves a question, just how in one day do you recreate an experiment to produce a primordial soup which originally took several weeks? Well, that is the Magic of television.

Chapter 23

A Trainee Scientific Glassblower

Paul Le Pinnet

Fellow of the British Society of Scientific Glassblowers

The day of his interview, Mr Murray had to report to "Reception" at the S.O.G. Laboratory complex, Runcorn Heath, Cheshire. The receptionist was someone I knew to be very efficient if not intimidating but quite nice when you got to know her. Sometime later Phil did admit to being a little intimidated and completely out of his depth.

My first impressions on meeting Phil were that he was large, soft spoken and respectful. My colleague and I took him to the interview room, we found that he had good sound academic qualifications and that he could express himself well. After a while we presented him with a Davis Double surface condenser and asked him to draw it to test his observational skills and to some extent his technical drawing ability. He drew a reasonable representation; he had seen that there were three layers of glass and was not bemused by the shiny surface.

We withdrew to the glassblowing workshop where he was shown the various items we made, and the equipment used.

At the bench, I demonstrated the type of flame required from the burner, then taking a length of pre-cut tubing I explained the hand positions, adjusted the flame and pulled a spear point while explaining how the hot glass was reacting and how I was adjusting to the softening glass and the radiated heat from both the flame and the hot glass, I then pulled a second spear point, turned down the flame and we changed places.

I watched carefully as he re-adjusted the flame, I corrected his hand positions a little and talked him through the process two or three times, I then stood back to observe from a little further away. There is a natural tendency at first to withdraw the glass from the flame as soon as it starts to soften, I explained that he had to keep the glass in the flame to "to allow the glass to absorb the heat" and so had to keep rotating the glass to enable him to control it, then once out of the flame keep rotating while the outer surface cools slightly before pulling apart while constantly rotating to produce a strong straight spear point.

There is a vast amount of information to assimilate, not only the hand–eye co-ordination but having this solid material changing to a liquid while being expected to control it as your delicate little hands are gradually getting hotter!

I watched carefully as he concentrated fully on what he was being asked to do, I thought to myself I could teach this young man. The interview ended with tea, biscuits and a chat.

There were other applicants to see, they were given the same interview, my colleague and I discussed at length their various merits.

The final decision was to invite Phil Murray back as a trainee Scientific Glassblower, which turned out to be a better than expected decision. On his first day, Phil was bright enough to turn up wearing a shirt and tie (which my colleague and I always wore) as we daily had face to face meetings with Chemists, Research Scientists and Managers, it was all based upon respect.

On the first day, I took him to meet key members of staff, i.e. the canteen ladies, then the man in charge of the stores and goods inward were we ordered lab coats and personal protection equipment, then on to the Finance controller, Safety department staff, Personnel department and finally our immediate Manager, which for him you could say was a fait accompli.

Phil Murray's sole role was to learn, which he did conscientiously to my relief and delight, his enthusiasm to learn something new each day never diminished for the seven years he was with me. He was always interested to see how we dealt with various customers, what questions to ask, our approach and the advice given. Even when he was working at the bench and I was dealing with a customer, I knew he was listening!

He was enrolled into the British Society of Scientific Glassblowers and was encouraged to take part in the various trainee competitions and attend the annual Symposia.

Our Engineering Manager offered to take all three of the glass-blowers to a production company in York to give Phil a little more experience. They were after all our main supplier of rod, tubing, taps, joints and flasks. On the way, we stopped off for breakfast. I noticed that there was a vegetarian option which included Linda McCartney vegetarian sausages! I was going on and on about just how nice these sausages were. Phil ordered the vegetarian breakfast with the addition of three rashers of bacon and three black puddings. The waitress who was a lady of a certain age smiled gently and said, "bless he's a growing lad". We continued to the supplier, it did Phil good to see a production line and to meet the lady at the other end of the telephone when we ordered joints and tubing.

I did on one occasion take advantage of Phil's six foot eight-inch height. The engineering workshops were just down the corridor, one day the fifteen occupants were in a particularly playful if not rowdy mood and were being quite noisy, I said to Phil "come with me and just stand behind me" I walked into the engineering workshop knowing Phil was towering over me and said "keep the noise down there are people trying to sleep". I asked, "any questions"? I knew that they were looking past me at Phil who they had seen duck down to get through the door, all went quiet until someone said "fine OK" they were all smiling their little faces off as we left.

Phil Murray by his own volition and enterprise applied for and succeeded in attaining a Churchill Fellowship, which enabled him to travel first to Europe and then to the U.S.A. to meet other Scientific Glassblowers, which extended his skills and experience immensely. He stayed with glassblowers and their families, some of whom are contributors to this book. All the feedback I had from this time about Phil Murray has been heart-warmingly positive.

It was a delight and a pleasure to train him.

Figure 1. Phil Murray and myself outside the Guild Hall, London.

The most amazing gesture that Phil Murray bestowed on both myself and my wife Michelle was to ask if we would go with him to the Guild Hall London for his Churchill Fellowship presented to him by HRH the Duke of Kent (see Figure 1).

It was such an honor to be with him on this special day.

Chapter 24

On the Backburner: Resuming a Career in Glassblowing After a 15-Year Hiatus

Introduction

At age 25 I left glassblowing after investing a short but intense seven years of labor, attention and commitment to learning my craft and developing my practice. It was a difficult decision that left an uneasy feeling; not quite regret, I had a feeling with roots buried deep in nostalgia, reminiscence and a gratitude that someone saw potential in me and helped me realize it.

I originally stepped away from the craft at a juncture in my life, considering two options — go to University and study, or take a new job working in export markets (Africa, India, Middle East) for a company that manufactured glass related products. I chose the latter and spent the next 10 years travelling the world.

My only regret is that I have waited too long to return and didn't find some way to at least maintain my practice in between doing other things. My only advice is: Once you have started, do not stop. Though craft, knowledge and skills were only a part of what I gained from my short career in glassblowing, it shaped my behaviors and approach to problems in ways that at the time I did not understand.

I have carried these learnings and lessons with me throughout my life and adapted them when I've faced challenges and shared them in one form or another when I've encountered those others who are also trying hard and eager to learn.

Perseverance has been the most valuable skill/behaviour I took from my glassblowing apprenticeship. It has stood me in good steam to maintain the philosophy required to repeatedly and successfully overcome the cognitive obstacles that one faces when attempting to progress beyond the difficulty of working glass as a medium, and the inevitable mistakes and frustrations one experiences in developing their glassblowing practice.

Now, turning 40, I set my sights on becoming a glassblower again, not with thoughts of what could have been, but with an idea of what could be the makings of perhaps a 30-year career ahead of me. Not quite a hobby, not yet a part-time occupation, my journey of relearning and adapting my practice to homelife and online economy has not been easy. Maybe this could help someone else who might be trying to start glassblowing.

The Renaissance

During my training, the number of UK-based Scientific Glassblowers was rapidly in decline. Glassblowers retried, unreplaced, with Vultures circling anywhere a lathe or other equipment might become available. I was one of a handful of apprentices (by which I mean — I think — there were about three of us in the country), which had its benefits as I was outnumbered by skilled people who were keen to pass on their knowledge and methods.

When I returned to the craft, I began searching for other lampworkers**, partly fearful that there would be hardly any left practicing.

It seems in my time away from the craft, lampworking as a discipline has undergone a renaissance.

Largely driven by the market for smoker's paraphernalia, lampworkers seem to have surged in numbers, the phenomenon birthing a creativity that blends sophisticated glass artistry with solid basic

---

**Throughout writing this piece, I've used the words lampworking & glassblowing interchangeably, though I refer to lampworking as the practice of fusing glass using a torch or flame, distinct from traditional furnace glassblowing.

scientific glassblowing principles. Discovering this after being away from the craft has been nothing short of wonderful.

The counter-culture element has driven other developments not just in the craft of shaping glass but in the industries associated with it. The availability of coloured borosilicate glasses is incredible compared to when I started out, and the corresponding techniques and artistry involved in not only shaping the glass but producing the desired uniformity of colour (or indeed variation in colour). Tooling has advanced massively, and crossovers in the glassblowing disciplines have emerged where lampworkers are using tools more typically associated with furnace glassblowing, such as glass jacks and rod yokes.

Nonetheless, these developments in tools and materials are really nothing in comparison to what I consider to be a real treasure in this re-invigoration of the craft — the community of glassblowers.

Unsurprisingly (considering the way it has changed the world), communication and media technology has in my opinion played the biggest part in helping the craft thrive. The process and experience of knowledge transfer has benefitted hugely from this, as well as making it easier for glassblowers to "monetize" their skills and sell what they produce on various online marketplaces.

Early in my career I was fortunate to be awarded a traveling fellowship from the Winston Churchill Memorial Trust, which enabled me to travel to the USA and Germany and learn from highly regarded Scientific Glassblowers. This was a life changing experience, and to this day I am thankful to those who helped me. The way technology has changed this is that now one can learn from others without leaving the home or workshop.

Social media platforms allow people to share their art, craft and methods across geographical boundaries, which I think has been a huge contributor to the lampworker culture that has emerged. I was fortunate to be part of such a culture in the past, which took the form of the societies of scientific glassblowers — particularly in the UK, Germany and the USA. It is amazing that, in the form of the networked accounts and pages on Instagram and Facebook, where (in my experience) ideas are shared and communication is open, there exists what is an informal, open-source network of collective expertise. Maybe these worlds can learn from each other.

The emergence of an "experience economy" has made it viable for glassblowers to earn money from facilitating for others the experience of lampworking glass. Participants can try it in a safe way that would otherwise not be accessible to them, given the expense in setting up a workshop and detailed knowledge that is needed.

This helps propagate the "gene pool" of practicing lampworkers — some get the bug and start themselves, or with their appreciation of the craft support other practicing lampworkers by collecting their art (and thus making the upkeep of maintaining a workshop more viable).

The participants range from hobbyists to full-time lampworkers. Remarkably, while many of the lampworkers are trying to earn a living from their craft, in general they are not greedy, secretive or exclusive about how they work and, in my experience, share their expertise and advice openly and freely. I am glad this hasn't changed.

A Series of Challenges

I had to overcome various challenges in re-entering the craft, and these were challenges I did not face when I started as an employed professional. Working for established workshops, chemical companies or universities affords the glassblower with infrastructure and resources that overcome a lot of what one need to do when setting up a workshop from scratch.

I think many glass-hobbyists and self-taught lampworkers must be innovative and committed to even begin melting glass by their own means in the first place. Here is to them.

Personally, I found at the outset, the practicalities of building a place to work safely and obtaining tools with which to work dominated the agenda, but this gave way to other challenges when developing and sustaining my practice.

I put this simply as moving from getting the means to start toward building the means to keep going, and I have written about it in the order which I experienced it.

A Place to Work

The first challenge I had to meet was finding a suitable building in which to work in the domestic setting. I applied three little pigs thinking. A wooden shed is a bad choice. I built a concrete panel garage I obtained for free (I just had to move it) with a foil lined/heat shielded area in the roof above my burner. And for ventilation I installed plate fan vents and a high flow $750\,\mathrm{m^3/hr}$ cooker hood salvaged from a skip in the area above me.

The work area needs to be constructed of materials that do not easily burn (obvious right?). It is naïve to think one will maintain enough attention to make sure a wooden table will not catch fire. I defy anybody to resist the tunnel-vision effect that occurs when one is anticipating thermal shock on a complicated workpiece.

Finally, a safe, ventilated secure area was required to store cylinders of fuel and oxygen.

Setting up a Space to Work in

Test runs with my burner confirmed which planes of space were most important. For safety, I found ceiling height and reach from the burner as important as lateral space. This was about heat throw – heat is projected forward from the burner and rises, so accounting for this depends on the burner type and how far the flame extends from the burner. Lateral space still needed to be adequate to allow un-cramped working. A glass tube is 5 feet in length, so at least that degree of space either side of the burner was my ideal. I decided if less space was available, then tubes can be halved before working (which will make them easier to work with than trying to relearn one's skills).

I setup a good 6 feet of clear space to the left of the burner and around 3–4 feet of uncluttered space to the right, to enable an arm to be outstretched, and a suitable headspace, to allow for hot gases rising from the burner – at least 5 feet and intumescent material above — I use aluminium plate on stand-off bushes to give an insulating air gap.

Ventilation was the next most important factor — allowing for both flow of air in and the extraction of fumes out at the workspace. Push–pull fan systems work well (extract air at the workspace, push air in from outside the workspace). Nitrogen oxides form on the fringes of torch-flames in oxy-propane combustion, and this gas is bad for the body, acidifying membranes in the eyes and lungs on prolonged exposure.

Burner & Gas Handling Equipment

I found that nowadays good burners are harder to come by cheaply, but new burners have never been easier to obtain. I found a wealth of information online, and again, a variety of designs are available on the market, I presume this has been helped by the resurgence in lampworking and the popularity of disciplines like beadmaking. My current personal preference is for simple, single valve controls for each gas. I use an old Jencons Rotojet burner because I already own it, and It will run on pretty much any fuel gas. It's not easy to use, but it gives a huge array of flame shapes and is extremely frugal. I use bottled propane for convenience. Future plans include a natural gas supply to my workshop, which will enable me to use a more versatile Herbert-Arnold Zenith burner I have in my possession. I found two-stage oxygen regulators easy to obtain. Two-stage propane regulators were not so easy to get so I adapted a solution to my purpose.

Fuel gas and oxygen are the big ongoing costs, perhaps more so than materials. Money can be saved negotiating cylinder rental between suppliers. Some suppliers place restrictions on collection methods — requiring a bulk-headed vehicle, trailer or pickup, i.e., they will not allow a normal family car to be loaded. I found alternatives, but this is worth knowing.

Oxygen concentrator units can be bought as an alternative to oxygen cylinders, but I'm informed that without sophisticated compressor/storage setups their output is only suitable for working smaller pieces like beads or marbles.

The UK HSE publishes a comprehensive, free guide to the safe use of welding gases (coded HSG 139) at the time of writing and is easily searchable online. This is a must to download and read. Gas companies like BOC and others also publish safety information on their websites. There is no better investment in tools than new, good quality gas regulators, flame arrestors and non-return valves.

The Kiln/Annealing Oven and Polariscope

To me this was almost more important than obtaining a burner, much in the same way that, in a fast car powerful brakes are probably more important than a powerful engine. While flame annealing can be used as an interim measure for annealing one's work, it is time- and fuel-consuming. At the outset I sourced a suitable annealing kiln, paying attention to find one of around 3 Kw, single-phase powered. This means that it can run from a standard 13 amp domestic socket. Pottery kilns are fine and can be run at a lower temperature (which will extend their life a great deal). Kiln controllers can be almost as expensive as a small used kiln, and for small-scale work, I have found sophisticated soak/ramp programs are not necessary and a simple temperature soak/cut-off module has been suitable for smaller items that fit into a $30 \times 30 \times 30$ cm kiln chamber.

In terms of economy, choosing high quality/high wattage timer switch and a domestic energy plan for Electric Vehicles was a money saver (they give a heavily discounted rate between 01:00–06:00 hrs).

A strain polariscope was another item I deemed essential — essential enough to store it for 15 years and carry it from house move to house move until the day I would use it again. I see this ability to qualitatively check your work for residual strain is a must for any work that ventures into technical or utilitarian purposes (i.e. anything other than non-functional art). It was also useful as a qualitative check of my skills (in flame annealing and strain balancing). Again, expensive though easily obtainable through a quick internet search.

Tools of the Trade

Most of those who work materials have an affinity for good tools. Interestingly, I found not making tools right at the outset helped my skills return, using gravity and push–pull techniques to shape the glass. To take my practice further though, it became clear that I would again need tools to physically push and pull the glass into shape.

Personally, I try to make my own and buy materials to build them rather than buy finished tools, because it brings me closer to the material, though where machining is required (e.g., Marble moulds) it makes sense to buy them and the prices are generally reasonable. I found graphite rod and plate could be bought easily online to make reamers and plates. Stainless and Brass rods and plates have been useful and cheap for making sharper edged reamers, with simple beeswax or high-end synthetic motor oil used as a lubricant (e.g., Mobil 1).

It is important to keep the lubricants away from any oxygen handling equipment and fittings when not being used.

Dressing for the Occasion

I omitted to buy myself a sturdy cotton lab-coat or coverall before starting up work and so lost some old but treasured items of clothing to minor scorches and burns. Massively frustrated with myself, I realized that I had let myself wander into complacency, which though mild and benign, demonstrated that I'd let my standards fall short at the outset. This resulted in avoidable mishaps that could have been worse. (even if they only affected me).

Reflecting on the past, I realized that I learned this lesson on my first day at work when my apprentice master (the author of this book), glued a paper tie to my t-shirt because I had turned up for work without a collared shirt and necktie. I could recall the lesson, but I had forgotten about its meaning. Naturally, because my workshop is at home, I would work in whatever I was wearing that day. My apprentice master got wind of this mishap through social media, and a few days later a parcel of neckties and a handwritten letter arrived at my house. Lesson relearned. Mindset — think professional, act professional and work professional.

Working from Home

The has been the biggest challenge to overcome. Unfortunate timing, I committed to dramatically increase the time spent in my modest workshop just before the dawning of the corona virus epidemic. This was a busy time for me personally and professionally (in my main job). Throughout the year's various local lockdowns and social distancing, this did however afford some opportunity to divert leisure time into the glassblowing project that might otherwise have been spent elsewhere.

Unsurprisingly, I found the time invested was directly related to an improvement in skills (do more to get better). My apprentice master called this "practice makes average".

The learning I took from this has been to organize time to make the best of it. A home workshop might seem supremely accessible, until competing demands like family, other grown-up commitments and a busy day job mean there might as well be a desert or an ocean between you and the glassblowing bench. This really eats at you, whether practising glassblowing for a hobby or a business enterprise.

Ultimately, my key learning here was the realization that an hour scrolling glass related feeds on social media was an hour that could have been spent practicing (though it is an unbelievably valuable knowledge source). Organizing my time became a valuable exercise and helped maintain the discipline required to relearn my skills and maintain the required mindset. My evenings and weekends are planned so I can accommodate work at the bench.

Finding an Outlet

Finding a way to extract value from my practice has been important in developing my craft, and my equipment. This is not necessarily a mercenary approach to monetizing the workshop output, but it is useful in many ways to find an outlet for what one produces, if only to perpetuate ones need and appetite to make something else, more complicated, or just made a little bit better each time.

Selling online is an obvious way to do this, which can be hard at the start, more emotionally really — especially if nothing sells. I have been there, and persevered. I found meeting customer demand to be a hugely motivating force (once the demand is there), and it can be a very fulfilling experience provided the customers' expectations against one's ability to produce are managed in a transparent way. This was again an old lesson I learned in the art of saying no, in the right way.

In terms of finding and using these outlets for my work, I went about it in what I saw as four overlapping categories:

Gifting: The making and giving of simple items for friends, family, charity, from this, I benefitted from the skills of practice and the appreciation of others.

Gifting to request: Trying to make something a bit more complicated than someone asked for. From this I benefitted from the challenge and the appreciation, and sometimes something in return (the crafts of other makers). I found this category can grow quickly so it was important to limit it. It is a good way to feed into the more commercial stages of:

Selling: Simple items I could make without much difficulty but that required consistency in the outcomes and finish. These I pre-produced and listed online, like selling a catalogue. Making something that has already been sold and not produced can be stressful or can be healthy pressure depending on the complexity of the item, and one's mindset.

Bespoke Making (making to order): more advanced, making more complicated or abstract items by someone else's design, the end goal in some ways for those offering services as this requires confidence, competence.

Finding an outlet was a key motivator for me. It became a boundary space between a hobby and a craft which might turn into a profession. Profiting from it, even only a little, was important if only to ensure my work and effort was valued, (and importantly so as not to devalue the work of others) and to reinvest in improving what I am trying to do. The healthy challenge of accepting appropriate commissions is another factor that helps bring on one's skills and practice. I set myself the parameters of taking payment once completed — not before — and to only take commissions that do not require a significant material outlay – this reduced the stress levels and mitigated the risk of not being able to deliver.

What Next

Though reflective, this account is intended to be anything but a personal memoir or opinion piece, and more to share the insights and experience for those that might be looking to start up their own home workshops.

I am hopeful established glassblowers might read this and find a way to use somehow this perspective of moving in and out of the craft, if not in their own endeavors, perhaps in supporting someone else's. I wonder what is next for scientific glassblowing — and I am hopeful that the popularity and practice of the craft will continue to widen. Most likely, this will be in new ways, driven by the connectivity that we can all have now.

Social media provides a community connection for those practising the craft, to share what they know and what they make. Streaming media with shows like "Blown away" is piquing public interest in glassblowing and the drama goes with it (by which I mean the drama in the challenges posed by glass as a medium, not the dramatics of being a glassblower).

The emergence of craft-related micro-economies, online marketplaces, and craft-entrepreneurship has made the income from such activities valuable to the lone craftworker and sometimes sustainable as a sole income source. Many lampworkers who are online sellers are one-person workshops, making to sell (and I hope thriving).

Modern connectivity means that artists and glassblowers can and do collaborate over distance. Collaborators need not be in the same country let alone in a workshop together. Often the lampworking videos posted online are of a solitary glassblower, self-filmed with a smartphone (me included).

Personally, I miss the company of other glassblowers; the friction, support, competition and fraternity that comes from working alongside one another. Perhaps it's still there for me to rediscover, albeit in a more "virtual" context.

This is not new and not a sad thought at all, though I am still pondering what this might mean for Scientific Glassblowing. Does this mean that time at the torch for the scientific glassblower is a little more, or a little less, lonely? I recall many university glassblowers worked alone (I did) and perhaps metaphorically, Scientific Glassblowing has always been a lonely discipline; when the glass is hot and you are behind the torch, you are on your own. There are not many ways someone else can help.

Many producing craft and selling online want to succeed but do not want to "grow" to the extent that they need to employ staff or extend themselves beyond their own ability to produce.[1]

I wonder if this will mean that resurgence of lampworking as an industry might perhaps plateau at the level of lone artisans making to sell. I use the word plateau without any detriment or insinuation of lacklustre, and this does not infer that the craft will suffer from knowledge stagnation.

When I entered the profession and started my apprenticeship, I had dozens of conversations with glassblowers reflecting on their approaching (and often past-due) retirement and lamenting the lack of support for apprenticeships. "Nobody is training glassblowers," they said. At the time they were right. What has changed is, it seems, is that glassblowers have begun training themselves from an alternative or zero skill base.

I have encountered self-taught lampworkers who just started up, driven by their enthusiasm for the medium out of an interest in smoker's paraphernalia or beadmaking. Some have taken short courses or experience days and gone from there. Graduate artists are adopting and adapting lampworking skills and techniques into their art. I have talked to furnace glassblowers that rent workshop time, who want to learn lampworking as a more accessible means to make their art or produce their wares.

---

[1] https://extfiles.etsy.com/Press/reports/Etsy_NewFaceofCreativeEntrepreneurship_2015.pdf.

So far, I have used the terms lampworker and glassblower interchangeably, but in summing up I need to narrow the definition to that of the scientific glassblower. A Scientific Glassblower is of course, a skilled lampworker, but the added knowledge of mechanical engineering principles and applied science is the key ingredient to providing a scientific glassblowing service — especially for supporting research. This too can be taught and learned.

Filling this gap is where I see the intrinsic value in publications such as this one. I hope awareness that this collective, recorded knowledge exists will filter into the informally networked communities of glassblowers, and that those with a bias toward the technical will develop their practice in the direction of scientific glassblowing.

The boundaries between craft, art, engineering and science become ever more blurred, not just conceptually but in the physical business of doing glassblowing, as a result of these converging economic, social and technological factors.

If industry and the community of enthusiasts can be mutually receptive to what could be a growing feedstock of potential candidates or collaborators, maybe a new paradigm for scientific glassblowing can be forged around the informal community of lampworkers and embraced by more formal associations to exploit the connectivity we have now.

Whatever keeps the craft alive.